SpringerBriefs in Water Science and Technology

More information about this series at http://www.springer.com/series/11214

Joao Mutondo · Stefano Farolfi
Ariel Dinar

Water Governance Decentralization in Sub-Saharan Africa

Between Myth and Reality

 Springer

Joao Mutondo
Eduardo Mondlane University
IWEGA
Maputo
Mozambique

Ariel Dinar
School of Public Policy
University of California
Riverside, CA
USA

Stefano Farolfi
UMR G-Eau
CIRAD
Montpellier
France

ISSN 2194-7244   ISSN 2194-7252   (electronic)
SpringerBriefs in Water Science and Technology
ISBN 978-3-319-29420-9   ISBN 978-3-319-29422-3   (eBook)
DOI 10.1007/978-3-319-29422-3

Library of Congress Control Number: 2016930823

This Springer imprint is published by SpringerNature
The registered company is Springer International Publishing AG Switzerland

Acknowledgments

This study is a result of the implementation of the Water Governance Decentralization in Africa: A Framework for Reform Process and Performance Analysis, project funded by the World Bank and the Water Research Commission of South Africa. It was undertaken by researchers stationed in the Water Science and Policy Center at the University of California, Riverside, and in the International Centre for Water Economics and Governance for Africa in Maputo, Mozambique. We would like to acknowledge individuals and organizations that were involved in the project which yield this study.

Special thanks are due to the World Bank and Water Research Commission of South Africa for the financial support, to Ariel Dinar for leading the scientific part of the project and sharing ideas, knowledge, documentation, and previous studies as well as the questionnaire that was used to analyze similar issues elsewhere and to Javier Ortiz Correa by his contribution in the quantitative data analysis, and to other project team members namely Stefano Farolfi and João Mutondo.

The primary data and information used in this study come from three case studies and a continent-wide survey. The three case studies were produced as M.Sc. dissertation, and therefore, we thank the authors of the case studies including their supervisors: Manuel Matsinhe (supervised by Eric Mungatana and Emilio Tostão at the University of Pretoria), Terence Chibwe (supervised by Magalie Bourblanc and Johann Kirsten at the University of Pretoria) and Gift Musinake (supervised by Vupenyu Dzingirai at the University of Zimbabwe). We also thank Derek Weston and Amos Mtsweni from Pegasys Strategy and Development, South Africa for the implementation of the African-wide survey.

This study would not have been possible without the responses of the river basin respondents, and therefore, we thank the representative of these river basins including the individuals who filled the questionnaire as well as the water agencies and other partners of the respective countries by facilitating the continent-wide data collection.

Contents

Acronyms and Abbreviations

ALG	Autorité de développement intégré de la région du Liptako-Gourma
ANJCC	Angolan Namibian Joint Commission of Cooperation
ARA	Administração Regional de Águas
BLIS	Baixo Limpopo Irrigation Scheme
CENOE	Emergency Operative Centre
CICOS	International Commission of Congo-Oubangui-Sangha
CMA	Catchment Management Agency
CMC	Catchment Management Committee
CRA	Conselho de Regulação de Águas
DNA	National Directorate of Water
DWAF	Department of Water Affairs and Forestry
FRELIMO	Frente de Libertação de Moçambique
HICEP	Chokwe Hydraulic Public Enterprise
ICMA	Inkomati Catchment Management Agency
INGC	Instituto Nacional de Gestão das Calamidades
IWMA	Inkomati Water Management Area
IMF	International Monetary Fund
IWRM	Integrated Water Resource Management
JIA	Joint Irrigation Authority
JWCSSA	Joint Water Commission-Swaziland and South Africa
KOBWA	Komati Basin Water Authority
LBPTC	Limpopo Basin Permanent Technical Committee
LCBC	Lake Chad Basin Commission
LHDA	Lesotho Highlands Development Authority
LIMCOM	Limpopo Watercourse Commission
LRC	Limpopo River Basin Commission
LVDP	Lake Victoria Development Programme
LVBC	The Lake Victoria Basin Commission
MICOA	Ministry of Environmental Coordination
MOPH	Ministry of Public Works and Housing

MRU	Mano River Union
NBI	Nile Basin Initiative
NGO	Non Governmental organization
NNJC	Nigeria-Niger Joint Commission for Co-operation
NWA	New Water Act
NWMP	National Water Master Plan
OERS	Organisation des Etats Riverains du Sénégal
OKACOM	The Permanent Okavango River Basin Commission
OMVG	Gambia River Basin Development Organization
OMVS	Organisation pour la Mise en Valeur du bassin du fleuve Senegal
ORASECOM	Orange-Senqu River Commission
PJTC	The Permanent Joint Technical Committee
PWC	Permanent Water Commission
RBO	River Basin Organization
RENAMO	Resistência Nacional de Moçambique
RNC	Niger River Commission
SADC	Southern African Development Community
SSA	Sub-Saharan Africa
TPTC	Tripartite Permanent Technical Commission
WUA	Water User Association
WMA	Water Management Area
UNECA	United Nation Economic Commission for Africa
UGBL	Unidade de Gestão da Bacia de Limpopo
ZAMCOM	Zambezi Watercourse Commission
ZANU	Zimbabwe African National Union
ZAPU	Zimbabwe African People's Union
ZINWA	Zimbabwe National Water Authority
ZRA	Zambezi River Authority

Chapter 1
Introduction, Motivation and Approach

Abstract The decentralization of water resource management at appropriate level has been subject of debate. The concept gained acceptance in 1992 in Dublin after the international conference on water and environment. As a consequence of the acceptance, most African countries have been implemented decentralization of water resource management through using the integrated water resource management framework. However, the factors affecting the level of decentralization process and performance of water resource management are still unknown and not largely studied in African context. Therefore, this chapter highlights the driving factors yielding the conduction of the present studies, presents the objectives of the study, summarizes the approaches used to analyze the objectives and state how those approaches are different from those used in previous studies.

Keywords Decentralization process and performance · Integrated water resource management

1.1 Motivation of the Study

Among the four so-called Dublin principles (ICWE 1992) representing the pillars of the worldwide acknowledged concept of Integrated Water Resource Management (IWRM), stakeholders' participation is the one calling for the definition of river basin management at the lowest appropriated level. This refers to the concept of decentralization of water management and governance. Blomquist et al. (2005) indicate that effective decentralization requires devolution of authority and responsibility from the center, and acceptance of that authority and responsibility by local entities in the basin.

This concept gained acceptance after the International Conference on Water and Environment in Dublin in 1992, and was supported by neoliberal institutions including the World Bank (Gleik 2002). For example, the World Bank Board took the lead by endorsing a water resource management policy paper that outlined a policy and strategy for Integrated Water Resource Management (IWRM), which has been used as the basis for water resource management throughout the world (World Bank 1993a). In Southern Africa, like in most developing countries, the World Bank supported the same reforms. To this effect, the World Bank coordinated a regional water resource management workshop at Victoria Falls in Zimbabwe, which detailed how to plan integrated water resource management (World Bank 1993b).

As a consequence of the acceptance of the IWRM, most African countries voted their water laws in the past fifteen years, and restructured their institutional and governance framework accordingly. For instance, South Africa instituted its national water act in 1998 followed by its national water resources strategy in 2002. Zimbabwe passed its water act in 1998; Zambia amended in 1994 its water act of 1970, while Mozambique and Tanzania approved their national water policies in 1995 and in 2002, respectively; Namibia passed its water resource management act in 2004.

The African governments have been implementing the approved water laws and policies. In Mozambique, the first national water law created five regional water administration agencies (ARAs) to implement integrated water resource management at the river basin level across the country. In Zimbabwe, the new water act established catchment and sub-catchment councils to manage seven major watersheds identified in the country in an attempt to decentralize the water management. South Africa's 1998 water act established nineteen water management areas (WMAs). Within each WMA, the law established the progressive creation of Catchment Management Agencies (CMAs), sub-catchments entities (Catchment Management Committees—CMCs) and water user associations (WUAs).

While much effort and goodwill were put into decentralization reforms in many basins in the continent, results have not been uniformly realized. For instance, the benefits originated from the implementation of such decentralization processes were taken for granted during the design of the South Africa National Water Act. However, ten years after the launch of the new national water policy, only two CMAs have been established and are operational (Inkomati and Olifants-Doorns). Moreover, many WUAs still struggle to find their place and role in the complex and sometimes confused context of water management in South Africa, while the Catchment Management Committees (CMCs)[1] exist only on paper, as they are *forums* with no decisional power. In other African countries, the process of decentralization in the water management institutions is even less advanced than in South Africa.

In Mozambique, for instance, in the early 1990s the water sector was highly centralized with all planning, implementation, operational responsibilities, and functions at the central level were performed by the National Directorate for Water. With the approval of the new water law (1991), the sector implemented comprehensive decentralization reforms by progressively setting up ARAs. The only ARA currently fully operational is ARA-Sul (South). ARA-Sul is responsible for the southern part of the country. As for the other regional water authorities, ARA Centro is already functioning, but needs continuing support, and ARA Zambezi is newly established. ARA Centro-Norte and ARA Norte have not yet been established.

---

[1]CMCs were supposed to create a link between the CMAs and the WUAs.

In Zimbabwe, in 1998 the government promulgated its Water Act and the Zimbabwe National Water Authority (ZINWA) Act. The new water acts established the creation of catchment and sub-catchment councils to manage seven major watersheds in an endeavor to decentralize water management. According to Musinake (2011), after almost fifteen years from its start, water decentralization in Zimbabwe is less likely to make a dent on livelihoods. For catchment communities to realize any meaningful benefits of decentralization and participation, it is critical that legislation, which includes both acts and the statutory instruments to mediate decentralization, is revisited and perfected.

In terms of water service provision, in Tanzania there is a broad consensus that the decentralization efforts through the local government reform program (LGRP) have brought better services closer to the poor people (e.g., access to rural water supply has increased from 43 % in 1990 to 53 % in 2005). Yet, the deficiencies in quantity and quality of services at local levels are still enormous (Egli and Zürcher 2007).

The process of water management decentralization in African countries is seen as a means of advancing river basin management at the lowest appropriate level. Although efforts in this direction are clearly identifiable in the continent, the very different stage of advancement in the African river basins' agencies witnesses the difficulty of implementing decentralization in practice.

It seemed necessary in this context to understand why some water agencies have succeeded more than others, what variables are involved in such reform process, which variables have a positive or a negative impact on the implementation of decentralization processes in the African water sector, and which variables could be affected by policy interventions and how. For this purpose, the specific objectives of the study are: (i) Analyze the factors that have potentially affected the results of decentralization process in Sub-Saharan African (SSA) basins, and (ii) Analyze the performance of the decentralization process in SSA basins. To answer the above objectives, this report uses and adapts to the local context an analytical framework developed for the same purposes elsewhere (Kemper et al. 2007). Different from past studies, the analytical and empirical framework developed in this report includes impacts of climate change and the transboundary nature of SSA basins in the analyses.

Past studies on the water management decentralization process have not fully addressed these issues. Kemper et al. (2007) initiated an investigation that was aimed at understanding the reasoning for (a) initiation of the decentralization process, and (b) variability in both initiation efforts and success of the decentralization process.

The general review in Mody (2004) yielded several important conclusions and research implications regarding the usefulness of comparative analyses of river basin decentralization processes. First, decentralization is a long-term process. It may fail at any stage and take turns, subject to internal and external shocks. Therefore, a snapshot of the decentralization process could be misleading in its comparative success or failure. Second, each river basin is a special case that cannot be compared to other basins. Therefore, the decentralization process that was designed to address conditions in one basin may not be relevant to another. Third, conclusions based on this case study approach should not imply that due to the unique conditions in a given basin it should be excluded from becoming a potentially good candidate for learning from extrapolating its experience to other basins. And finally, any type of cooperation among the various parties involved in the management of the basin water and other resources is a predictor for a stable and successful decentralization process (Blomquist et al. 2007: 229–238).

The comprehensive work by Kemper et al. (2007) has not addressed several important aspects. First, their analyses, both the case studies and the econometric ones, did not include basins from Sub Saharan Africa (SSA). Due to that exclusion, they omitted some important aspects of the decentralization process that, while being unique to SSA, are very much relevant to other continents. International river basins in SSA cover more than 60 % of land territory. Having a local basin nested in a transboundary basin arrangement (which some basins in SSA are subject to) would suggest additional explanation to success or failure of decentralization. We will expand on this aspect in Chap. 7.

And second, Kemper et al. (2007) conducted their analyses in isolation from likely climate change impacts on the water cycle. Climate change affects the inter- and intra-annual variability of water flow. SSA is one of the most climate-change-prone regions (Alavian et al. 2009). Dinar et al. (2010, 2011) claim that many basin-level management decisions are made with future perspectives in mind. They identified an inverted U shape of the likelihood for basin cooperation with regard to water scarcity and flow variability. We build on their analysis to establish several hypotheses with regard to the need, speed, and likely success of decentralization in SSA, and with regard to climate change and precipitation/flow variability. It is this set of considerations that this study will address, departing from the analytical framework in Dinar et al. (2007).

No quantitative analytical framework to understand the factors of success and failure of decentralized water governance similar to the one adopted in this study has been applied to African catchments previously. The only examples of quantitative analysis to study water decentralization processes found in the literature are a case study run in Ghana (IBRD/WB 2007), and a recent study (Gallego-Ayala and Juizo 2012) in Mozambique. The first study uses a network analysis, while the second uses quantitative synthetic indexes to assess the performance of river basin organizations to implement integrated water resource management. Several qualitative studies look at decentralization of water management and services in South Africa, particularly (Wijesekera and Sansom 2003; Chancellor 2006; and Zenani 2007), but no quantitative framework is proposed or applied so far.

1.2 Study Approach

In order to investigate the process and performance of river basin decentralization in Sub Saharan Africa, a two-tiered approach was developed in the study reported here. First, a detailed application of the case study approach in Blomquist et al.

(2008) was implemented in three river basins in Southern Africa (first phase). While the case study analyses highlight the direction in decentralization of river basin management, they do not permit the identification of generic reasons and forces behind the decentralization process and its performance. Thus a quantitative analysis of basins in SSA took place (second phase) based on the same analytical framework. This SSA study allowed an analysis of determinants of the water decentralization process and performance in the Continent.

1.3 Scope of the Study

This study is composed of seven chapters. The current chapter (introduction) is followed by a literature review (Chap. 2) about decentralization of water management in Sub-Saharan Africa, and elsewhere. The studies reviewed in Chap. 2 help to identify the main challenges of implementing IWRM in Africa. Among them, the most relevant are the lack of clarity in terms of power relations; the insufficient financial sustainability of the managing agencies; the lack of knowledge and skills available to manage water at the various institutional and geographical scales; the conflicts raised as a consequence of increased decision-making power given to local actors with colliding interests; the unclear role of the state; the difficult public-private relations; the lack of reliable data and information; and the cultural impediments.

Chapter 3 describes the analytical framework used both in the case studies and in the quantitative analysis. Chapter 4 compares the results obtained in the three case studies and draws the status of the decentralization process and performance in the studied catchments. Chapter 5 presents the empirical models used to analyze the determinants of the river basin management decentralization process and performance. Different measures of decentralization process and performance are described, including the description of key variables affecting the decentralization process and performance, as well as their expected direction. Chapter 6 presents the results of the quantitative analysis, and Chap. 7 compares the results from the two approaches (qualitative vs. quantitative) and draws conclusions and policy recommendations.

References

Alavian, V., Quddumi, H. M., Dickson, E., Diez, S. M., Danilenko, A. V., Hirji, R., et al. (2009). *Water and climate change: Understanding the risks and making climate-smart investment decisions*. Washington DC: World Bank.

Blomquist, W., Dinar, A., & Kemper, K. (2007). River basin management: Conclusions and implications. In Kemper et al. *Integrated river basin management through decentralization* (pp. 229–238). World Bank and Springer.

Blomquist, W., Dinar, A., & Kemper, K. (2005). Comparison of institutional arrangements for river basin management in eight basins. World Bank Policy Research Working Paper #3636. Washington, DC.

Blomquist, W., Dinar, A., & Kemper, K. (2008). A framework for institutional analysis of decentralization reforms in natural resource management. *Society and Natural Resources. 23*, 1–16. Routledge: Taylor & Francis Group.

Chancellor, F. (2006). Crafting water institutions for people and their businesses: Exploring the possibilities in Limpopo. In Perret-Farolfi-Hassan (eds.) *Water governance for sustainable development*. London: Earthscan.

Dinar, A., Blankespoor, B., Dinar, S., & Kurukulasuriya, P. (2010). Does precipitation and runoff variability affect treaty cooperation between states sharing international bilateral rivers? *Ecological Economics, 69*, 2568–2581.

Dinar, A., Kemper, K., Blomquist, W., & Kurukulasuriya, P. (2007). Whitewater: Process and performance of decentralization reform of river basin water resource management. *Journal of Policy Modeling, 29*(6), 851–867.

Dinar, S., Dinar, A., & Kurukulasuriya, P. (2011). Scarcity and cooperation along international rivers: An empirical assessment of bilateral treaties. *International Studies Quarterly, 55*, 809–833.

Egli, W., & Zürcker, D. (2007). The role of civil society in decentralization and alleviating poverty: An exploratory case study from Tanzania, ETH-NADEL Report, 31p + Ann.

Gallego-Ayala, J., & Juízo, D. (2012). Performance evaluation of river basin organizations to implement integrated water resources management using composite indexes. *Physics and Chemistry of the Earth, 50–52*, 205–216.

Gleik, P. (2002). *The world's water: The Biennial report on freshwater resources*. Washington: Island Press.

ICWE. (1992). The Dublin statement and report of the conference. In *International Conference on Water and the Environment*, January 26–31, 1992.

IBRD, & World Bank. (2007). *Tools for institutional, political, and social analysis of policy reform* (pp. 370–379). Washington: International Encyclopaedia of the Social Sciences 2. Mac Millan. New York.

Kemper, K. E., Blomquist, W., & Dinar, A. (2007). *Integrated river basin management through decentralization*. World Bank and Springer.

Mody, J. (2004). Achieving accountability through decentralization: Lessons for integrated river basin management. World bank Policy Research Working Paper 3346. Washington DC: World Bank.

Musinake, G. (2011). Reform process and performance analysis in water governance and decentralization: A case of Mzingwane catchment in Zimbabwe. MSc. Thesis. CASS, University of Zimbabwe, Harare.

Wijesekera, S., & Sansom, K. (2003). Decentralization of water services in South Africa. In *29th WEDC International Conference on "Towards the Millennium Development Goals,"* Abuja, Nigeria.

World Bank. (1993a). *Water resources management*. Washington DC: World Bank.

World Bank. (1993b). *Proceedings at the Workshop on Water Resources Management in Southern Africa*. Zimbabwe: Victoria Falls.

Zenani, V (2007) *Institutional dimensions of water resource management in South Africa,* Water Research Commission Report: University of Cape Town.

Chapter 2
Water Decentralization Experiences: A Literature Review

Abstract As noted in Chap. 1, most of African countries have adopted integrated water resource management after the international conference on water and environment, which took place in Dublin in 1992. The implementation of the integrated water resource management has been different within the continent and even within the same region or country. This scenario has produced different results existing within the continent some good experiences while other countries and region are lagging behind. In order to understand the different forms of implementation of integrated water resource management, this chapter describes the process followed in different countries. The description is centered in the policies and other legal instrument enacted by the respective governments, and highlights the key success and failure histories and finally discuss in critical point of view the results obtained from different experiences during the implementation of integrated water resource management in Africa.

Keywords Decentralization process and performance · Integrated water resource management · Water policies

2.1 Background

Although the concept of decentralization has been attempted and practiced over decades, its application to water resources, especially in Sub-Saharan Africa, is contemporaneous and unprecedented. Water management decentralization reforms, based on the principles of Integrated Water Resource Management (IWRM) were characterized by several aspects. They established catchment and sub-catchment organizations, adding another layer of institutions to those dating back to the pre-independence or to the immediately post-independence frameworks. The reforms were presumed to redress problems of inequitable access, high pollution levels, seasonal scarcity, and ever-increasing conflicts. Such conflicts had bedeviled the water sector, as well as delivering water and livelihoods for the people, especially the poor, through incorporating them into the decision-making process.

© The International Bank for Reconstruction and Development/The World Bank 2016
J. Mutondo et al., *Water Governance Decentralization in Sub-Saharan Africa*,
SpringerBriefs in Water Science and Technology, DOI 10.1007/978-3-319-29422-3_2

Studies elsewhere show decentralization endeavors to be successful in some cases while unsuccessful in others. Dinar et al. (2005) recommend decentralization of water management by arguing that when decision-making is centralized and local conditions are not appropriately taken into account, then accountability of decision-makers is weak, and water resource management is inadequate. Empirical evidence from river basins in the developed and developing world shows that decentralization of water management has led to tremendous achievements in conflict and pollution reduction, productive and allocative efficiency, and environmental sustainability (Blomquist et al. 2005a, b, c, d; Dinar et al. 2005).

However, Stalgren (2006) argues that political entrepreneurs at the national level strategically position themselves by influencing the "construction of reality" in matters of water governance decentralization at the local level to their advantage. Smith (1983) and Fesler (1968) also point out that decentralization promotes parochial and separatist tendencies and may deepen enclaves of authoritarianism as well as exacerbate inequalities. Kambudzi (1997) states that democratization of water may go beyond our intention and turnout to be a recipe for further disaster.

In most Sub-Saharan African countries the level of awareness to the national reforms, as a starting point, differs significantly from country to country, catchment to catchment, sub-catchment to sub-catchment, and from locality to locality. Operations and effectiveness of the resultant institutional arrangements remain heterogeneous, even within the same national boundaries, in which laws and statutory arrangements governing the process are almost homogeneous. This fact suggests that the decentralization process appears not to be a linear and steady process in these countries. However, a thorough analysis of the factors that contribute to the success and failure of the water management decentralization process in these countries has not yet been conducted.

2.2   IWRM, Decentralization, and African Water Policies

At the heart of most of the water reforms that were implemented in Africa from the early 1990s is the concept of integrated water resource management (IWRM) (ICWE 1992). This concept is defined as "equitable access to and sustainable use of water resources by all stakeholders at catchment, regional, and international levels, while maintaining the characteristics and integrity of water resources at the catchment scale within agreed limits" (Pollard 2002, p. 943). The IWRM encapsulates each of the four Dublin Principles as follows (Swatuk 2005):

1. Fresh water is a finite and vulnerable resource, essential to sustain life, development, and the environments;
2. Water development and management should be based on a participatory approach, involving users, planners, and policy makers at all levels;
3. Women play a central part in the provision, management, and safeguarding of water;

4. Water has an economic value in all its competing uses and should be recognized as an economic good.

Among these four principles, stakeholders' participation is the one calling for the definition of river basin management at the lowest appropriate level. This refers to the idea of decentralization of water policies implementation. In other terms, following the subsidiary principle, the design and implementation of water management and allocation policies are transferred from the state to local institutions, which are supposed to have a better knowledge of the catchment functioning and where representatives of local water stakeholders are able to negotiate and decide jointly water management strategies and measures to be put in place. It is what Ostrom (1990) calls *collective action* in the management of common pool resources through the design by stakeholders themselves of the rules governing those resources.

At the same time, the Dublin Statement of 1992 demands a holistic approach to management of water resources, linking social and economic development with protection of natural ecosystems and also linking land and water uses across an entire catchment area of groundwater aquifer. According to Mody (2004, p. 8), "this holistic approach thus entails greater integration and centralized decision-making in certain dimensions, while competition for resources makes feasible and increases the desirability of decentralization and stakeholder participation."

In other terms, while centralization in the river helps achieve coordination of infrastructure, human resource development and the setting of general priorities for water allocation, water quality, and land use, decentralization can achieve efficiency gains through more effective delivery of services to users, and also through more prudent use of local resources and initiatives.

In terms of economic efficiency and institutional effectiveness of the water governance set-up, centralization can take advantage of economies of scale, internalize externalities and manage the hydrological interconnectedness, but it suffers from the disadvantage of bureaucratic cumbersomeness and slow response. Decentralization on the other side risks the danger of raising transaction costs and requires the pre-establishment of a property right system on the resources (Mody 2004, p. 10).

Mody (2004, p. 12) concludes that there is no generic recipe for the identification of the lowest appropriate level of management in a river basin. This appropriate level can correspond to the river basin authority that offers participation, or it may be a water user association that monitors, operates, and manages a small-scale irrigation system.

African states, and particularly those of the Southern Africa Development Community (SADC) region, are primarily "a collection of economically weak, primary commodity exporting, debt-distressed countries with unconsolidated democracies" (Swatuk 2005, p. 877). This fact has important consequences on the budgets and human resources capacities that SADC countries in Africa can put in place in order to implement in practice the IWRM principles that underpin their water policies. Two exceptions in the region are represented according to Swatuk (2005) by water reforms in South Africa and Namibia.

According to Swatuk (2005), the main difficulties in the implementation of IWRM policies in the SADC countries can be identified by the following aspects: *institutions*, due to the institutional inertia that pushes towards maintaining and adapting existing institutions rather than creating new (decentralized) ones proposed by IWRM; *finance*, due to the troubles in finding economic resources and the dependence on foreign donors; *conflict resolution*, due to the significant intra-basin (and, to a smaller extent in the region, inter-basin) competition for use of the limited water resource; and *information*, due to the lack of reliable and valid data and information about the state of the resource.

Van der Zaag (2004), quoted by Swatuk (2005 p. 878), suggested during the opening session of a SADC meeting that "perhaps the creation of wholly new institutions for water resources management was a mistake. Rather, the new institutions might be more effective if they were endowed with advisory powers only, and that more effort should be made to introduce IWRM practices into existing bureaucratic forms and procedures."

The particular and disadvantaged situation represented by most African countries requires a specific approach with regard to the implementation of the concept of IWRM through water policies, and especially when it comes to decentralization. The complex, expensive, and non-linear nature of decentralization, combined with the difficult socioeconomic and institutional conditions of African countries, seem to create dubious pre-conditions for the introduction of a suitable environment for decentralization policies. The following section provides an overview of African experiences in terms of implementation of policies directed towards IWRM and decentralization.

2.3 Success and Failure Stories from Africa, with Focus on SADC Countries

It seems useful to look into concrete examples of recent water policy implementation in Africa and observe the assessments that authors have for these institutional dynamics, in light of the problems raised in the previous section.

Following Swatuk (2005) who uses South Africa and Namibia as two positive exceptions in the region, we will start our overview from these two countries and will proceed towards Botswana, Zimbabwe, Tanzania, Mozambique, Mali, and Burkina Faso.

Brown (2010) explored the institutionalization of participatory water resource management in post-apartheid **South Africa**, analyzing the situation in one of the two (out of the nineteen originally foreseen) catchment management agencies (CMAs) currently fully operational in South Africa, the Incomati CMA. The author argues that participation in natural resource management, often coupled with moves for more local ownership of decision-making, is based, among other things, on assumptions about the role of the state and the transformation potential of institutional reforms. Brown (2010) concludes that, after empirical research in the Incomati water management area, there might be fundamental weaknesses in the participatory model and the underlying assumptions. The implemented approaches of decentralization may actually reinforce inequitable outcomes rather than achieving equity, efficiency, and sustainability in the use of water and other resources.

Brown (2010, p. 183) advocates in South Africa, as in all transitional countries, a reassessment of the role of the state, which should be reinforced, as it moves toward participatory governance to not render traditional hierarchical government intervention obsolete, but overall because a "laissez faire" approach to water participation and decentralization by the Department of Water Affairs and Forestry (DWAF) in Pretoria could have provided opportunities for existing powerful water users and vocal groups to co-opt processes and dominate the new organs of governance of CMAs and water users associations (WUAs).

Hossain and Helao (2008) presented some experiences from Northern **Namibia** and shed light on how the management and distribution of water resources have changed in independent Namibia, within the background of the government's decentralization efforts. The authors observe that Namibia continues to suffer from acute water shortage, recognizing that decentralization is not a monolithic concept, neither is it inherently positive or negative. They conclude that there is very little evidence that the liberal and commercial approach adopted by the Namibian government towards water resource management resulted in policies that are more responsive to the poor or indeed to citizens generally. According to Hossain and Helao (2008), local governments are familiar with local circumstances, therefore, they may be in the best position to more equitably distribute public resources and target poverty within their own jurisdictions. However, redistribution issues from richer to poorer areas must be the responsibility of central governments. In this statement, the authors agree with the thesis of Brown (2010) in terms of the role of the state. A reason for concern, according to Hossain and Helao (2008), is represented by the importance that private interests have in the public decision-making process: "By promoting participatory good governance, grassroots-based local government institutions like the Oshikuku village council can ensure public trust much more easily than the private corporations" (Hossain and Helao 2008, p. 210).

Botswana is a Southern African country regarded by many authors as a "success story" because of nearly four decades of unabated economic growth, multi-party democracy, conservative decision-making, and low levels of corruption (Swatuk and Rahm 2004). The country faces increasingly high water scarcity, due to the dramatic rise in water use of water resources. Local policy makers recognized that water supply is limited in this arid/semi-arid country and took deliberate steps to manage water demand. Botswana then devised a national water master plan (NWMP) and undertook a series of institutional and legal reforms throughout the 1990s so as to make water resource use more equitable, efficient, and sustainable (Swatuk and Rahm 2004).

In other words, IWRM once again drove the design and implementation of Botswana's water policy. But according to the authors, policy measures have had

limited impact on the practice, due to a number of socioeconomic and political challenges, identified in: *the character and pace of development* (focus on infrastructure
development in support of jobs with negative consequences, and externalities on the
environment and on the use of natural resources); *institutional overlap* (too many
actors decide about water management, with little coordination from the government); *cultural impediments* (no general belief that water will run out and a sense that
"government will provide"); *human resource capacity* (lack of data, information, and
expertise); *Power relations* (the continuing preference for new supply, despite stated
support for demand management, reflects the tension between international and
national networks of power).

Swatuk and Rahm (2004, p. 1363) conclude that the current surplus capital rein-
forces the belief that water can be acquired "somewhere": alternatively, technology
will provide, and then "somewhat ironically, this wealth inhibits rather than fosters
sustainable water resource management."

According to Mapedza and Geheb (2010), **Zimbabwe** emerged as a country with
one of the most progressive (on paper, at least) water reform processes within the
Southern African region. Decentralization was certainly a milestone of the water
reform in the country. The 1998 Water Act set up a decentralized water management
structure, based on seven catchment councils. More than a decade after, the authors
state that water reform in Zimbabwe was not simply a technical process, but "it is
clearly linked to issues of power, political connectedness, and gender, with fewer
women benefitting from the largely violent fast track land reform process" (Mapedza
and Geheb 2010, p. 525).

Similar to the arguments quoted above by Brown (2010) about South Africa, and
by Hossain and Helao (2008) about Namibia, Mapedza and Geheb (2010) state that
"Zimbabwe's water reform has negatively impacted the livelihoods of the poor,
whose position is weakened by a lack of resources....-...How the reform played it
out in Zimbabwe is a function of unequal power dynamics amongst the stakeholders
…mechanisms should pro-actively be put in place to tilt the power asymmetries in
favor of the poor people in Zimbabwe, who largely rely on informal and multiple
water uses…" (Mapedza and Geheb 2010, p. 525).

Dungumaro and Madulu (2003) make reference to three experiences from
irrigation projects in **Tanzania**, leading to very different outcomes to stress the
importance of community involvement and participation into any developmental
initiatives, including water-related projects.

In 2006, the government of Tanzania launched a national program to meet, by the
year 2015, the water sector targets set out in the Millennium Development Goals.
According to Giné and Perez-Fouguet (2008), there is evidence that the government
is promoting more sustained facilities, focusing on cost recovery and on "decentralization by devolution." But shortcomings exist, due principally to a number of factors
determining non-sustainability of the program. According to the authors, "decentralization to the lowest appropriate level is usually interpreted as the need for local
communities to assume responsibility for their water supply, while little attention
has been given to define responsibilities of sector-related institutions, nor to methods
for tracking their performance" (Giné and Perez-Fouguet 2008, p. 18). For the

authors, the main challenge is identifiable in the management of the systems and in their financial sustainability, once installed. Operation and maintenance costs should then be covered by water users. Other important challenges hindering the performance of water decentralization in Tanzania are identified in the need for additional external funding, the lack of strategic vision by local authorities, the lack of skills, the crucial need for technical support, and the lack of a supervision and monitoring system.

In the field of urban and peri-urban domestic water supply, Matsinhe et al. (2008) looked at the possible synergies derived from the public-private partnership in the provision of water services in Maputo, **Mozambique**. The authors present the critical situation of the Mozambican capital in terms of water service provision (only 40 % of households have an indwelling water source), while 38 % of the population is served by small-scale independent providers (SSIP). To secure and improve water service provision to the poorest and most disadvantaged households of the city, the authors advocate the legalization of SSIP and the decentralization of certain regulatory functions from the central regulatory body (CRA—Conselho de Regulação de Aguas) to the neighborhood level. The sustainability of peri-urban water services regulation, based on neighborhood water committees, requires that CRA and the municipalities formalize a system of payments of license and regulatory fees to ensure long-term functioning of institutions created for the purpose (Matsinhe et al. 2008).

Looking at Western African water reforms, the national water law in **Mali** was voted in 2002, and was followed by the 2006 National Water Policy, based on the principles of IWRM (Water Aid 2008). Water management in the country is under the responsibility of the Ministry of Mines, Energy and Water, and decentralization has taken place since 2002, when local authorities (*collectivités locales*) were legally mandated for economic, social, and cultural development. In the water sector, the local authorities in charge of water management and allocation are the municipalities. The report of WaterAid indicates that, although on the technical side the decentralization process showed solid advances, financial concerns are still hindering the dynamics proposed in the policy. Financial problems and lack of investment funds represent, according to WaterAid, the main constraint that seriously risks jeopardizing the whole decentralization process (Water Aid 2008). The institutional reform of water policies in **Burkina Faso** took place in three big phases and is closely connected with the IWRM agenda at the international scale (Petit and Baron 2009). In 1998, the government adopted the "Water Policy and Strategies" policy document. Three years later, in February 2001, a Water Framework Law was approved by the parliament. In 2003, an action plan for the integrated management of water resources (IWRMAP) was proposed and covers a period until 2015.

Within the IWRMAP, a decentralization process took place and, as in other countries of the West African region, encounters serious implementation difficulties. Nevertheless, according to the authors, "we can mention concrete achievements, even if numerous dysfunctions still remain. For instance, a water agency was created in March 2007 in the Nakambé Basin, and about 20 local water committees have been created" (Petit and Baron 2009, p. 56). The main limits identified by the authors with regard to the implementation of the IWRMAP in Burkina Faso include: (a) the gap between design and implementation of the water policy; (b) the lack of clarity and the subsequent conflict of competences and power in the water sector; and (c) the lack of coordination between the national and the local level. This last aspect is particularly relevant for the decentralization process, which "is experiencing difficulties of implementation because of a lack of delegation of competencies, and because of the limited funds allocated to local authorities in the water sector" (Petit and Baron 2009, p. 57).

2.4 Some Reflections Emerging from the Literature Review

IWRM is a complex and expensive process, and decentralization is a crucial component of IWRM. Sub-Saharan African countries suffer from chronic inefficiencies and gaps inherited from their recent past, and do not seem to represent a proper socioeconomic, political, and institutional environment for the fast and successful implementation of such policies in the water sector.

Following the IWRM principles and recipes, most African countries reformed their water policies starting from the early 1990s, and put much emphasis on decentralization processes and the creation of new agencies at the local level for water management and governance.

The experiences illustrated in the previous paragraph show that although progress is visible in the field of water policies implementation and decentralization processes, many challenges still exist. Substantial differences are observable around African countries, but even those nations indicated as good examples in the difficult path toward the practical application of IWRM principles in the real life, like South Africa, Namibia and Botswana, still face delays and difficulties in the implementation of water policies, with particular reference to decentralization.

The main challenges are represented by the lack of clarity in terms of power relations and distribution of competences between central and local institutions, and between old and new organizations; the insufficient budgets and the lack of financial sustainability of the managing agencies; the lack of knowledge and skills (human resources) available to manage water at the various institutional and geographical scales; the conflicts raising as a consequence of increased decision-making power given to local actors with colliding interests; the unclear role of the state in the more participatory and "democratic" arena represented by local water forums, users' associations and agencies; the difficult public-private relations and the issue of delegation/devolution of power to private actors for the management of a vital resource; the lack of reliable data and information available for a responsible and effective decision-making process; the cultural impediments to water pricing for the recovery of investment and O&M (operation & maintenance) costs, both for bulk water and for water services.

As described above, the level of decentralization process is heterogeneous among countries and even within the same national boundaries, in which laws and statutory arrangements governing the process are almost homogeneous. This fact suggests that the decentralization process appears not to be a linear and steady process in these countries. This scenario indicates that the decentralization process and performance is affected by diversified factors, and an assessment of these factors contributing to the decentralization process and performance of water resource management is essential.

This study applies an institutional framework presented by Dinar et al. (2007), Kemper et al. (2007) and Blomquist et al. (2010) in an early global study to appraise the factors contributing to the decentralization process and performance of water resource management in African river basins.

The institutional framework used to analyze the factors behind the successful or unsuccessful decentralization process and performance is described in the next chapter. This framework is applied at both the case study (catchment) and the regional levels. We introduced several modifications to the original framework in order to address issues pertaining to Sub-Saharan Africa. We will detail these modifications in the following chapters.

References

Blomquist, W., Ballestero, M., Bhat, A., & Kemper, K (2005a). Institutional and policy analysis of river basin management: The Tárcoles River basin, Costa Rica. World Bank Policy Research Working Paper #361, Washington, DC.

Blomquist, W., Haisman, B., Bhat, A., & Dinar, A. (2005b). Institutional and policy analysis of river basin management: The Murray Darling river basin Australia. World Bank Policy Research Working Paper #352, Washington, DC.

Blomquist, W., Calbick, K., & Dinar, A. (2005c). Institutional and policy analysis of river basin management: The Fraser River basin, Canada. World Bank Policy Research Working Paper # 3525, Washington DC.

Blomquist, W., Tonderski, A., & Dinar, A. (2005d). Institutional and policy analysis of river basin management: The Warata River basin, Poland. World Bank Policy Research Working Paper #352, Washington DC.

Blomquist, W., Dinar, A., & Kemper, K. (2010). A framework for institutional analysis of decentralization reforms in natural resource management. *Society & Natural Resources, 23*(7), 620–635.

Brown, J. (2010). Assuming too much? Participatory water resource governance in South Africa. *The Geographical Journal, 177*(2), 171–185.

Dinar, A., Kemper, K., Blomquist, W., Diez, M., Sine, G., & Fru, W. (2005). Decentralisation of river basin management: A global analysis. The World Bank (Policy Research Working Paper 3637). Washington, DC: The World Bank.

Dinar, A., Kemper, K., Blomquist, W., & Kurukulasuriya, P. (2007). Whitewater: Process and performance of decentralization reform of river basin water resource management. *Journal of Policy Modeling, 29*(6), 851–867.

Dungumaro, E. W., & Madulu, N. F. (2003). Public participation in integrated water resources management: the case of Tanzania. *Physics and Chemistry of the Earth, 28,* 1009–1014.

Fesler, J. W. (1968). Centralization and decentralization. In D.L Sills, (Ed.), *International encyclopaedia of the social sciences* (pp. 370–379) New York: Macmillan and Free Press.

Giné, R., & Perez-Fouguet, A. (2008). Sustainability assessment of national rural water supply program in Tanzania. *Natural Resources Forum, 32*, 327–342.

Hossain, F., & Helao, T. (2008). Local governance and water resource management: Experiences from Northern Namibia. *Public Administration and Development, 8*(3), 200–211.

ICWE. (1992). The Dublin Statement and Report of the Conference. International Conference on Water and the Environment, 26–31 January 1992.

Kambudzi, M. (1997). Water democracy or water development? A challenge in setting priorities. In B. Derman & C. Nhira (Eds.), *Towards reforming the institutional and legal basis of the water sector in Zimbabwe* (pp. 6–8). Harare: CASS.

Kemper, K. E., Blomquist, W., & Dinar, A. (2007). *Integrated river basin management through decentralization*, World Bank and Springer.

Mapedza, E., & Geheb, K. (2010). Power dynamics and water reform in the Zimbabwean context: Implication for the poor. *Water Policy, 12*, 517–527.

Matsinhe, N. P., Juizo, D., Macheve, B., & Dos Santos, C. (2008). Regulation of formal and informal water service providers in peri-urban areas of Maputo Mozambique. *Physics and Chemistry of the Earth, 33*, 841–849.

Mody, J. (2004). Achieving accountability through decentralization: Lessons for integrated river basin management. World bank Policy Research Working Paper 3346, Washington DC: World Bank.

Ostrom, E. (1990). *Governing the commons: The evolution of institutions for collective action*, Cambridge, p. 280.

Petit, O., & Baron, C. (2009). Integrated water resources management: From general principles to its implementation by the state. The case of Burkina Faso. *Natural Resources Forum, 33*, 49–59.

Pollard, S. (2002). Operationalising the new Water Act: contributions from the Save the Sand Project-an integrated catchment management initiative. *Physics and Chemistry of the Earth, Parts A/B/C, 27*(11–22), 941–948.

Smith, B. C. (1983). *Decentralization: The territorial dimensions of the state*. Boston: M.A George and Unwin.

Stalgren, P. (2006). *Worlds of water: Worlds apart. How targeted domestic actors transform international regimes*. Göteborg Studies in Politics, Goteborg, Sweden, p. 228.

Swatuk, L. A. (2005). Political challenges to implementing IWRM in Southern Africa. *Physics and Chemistry of the Earth, 30*, 872–880.

Swatuk, L. A., & Rahm, D. (2004). Integrating policy, disintegrating practice: water resources management in Botswana. *Physics and Chemistry of the Earth, 29*, 1357–1364.

Van der Zaag, P. (2004). Integrated Water Resources Management: Relevant concept or irrelevant buzzword or keynote speech, 5th annual meeting of the WaterNET/Warfsa, Windhoek, Namibia, 2–4 November.

Water Aid. (2008). Mali, pour un financement local efficace des services d'eau et d'assainissement, *Water Aid Report*, p. 19.

Chapter 3
Analytical Framework

Abstract As illustrated in Chap. 1, this study aims to explain the factors affecting the level of decentralization process and performance of water resource management in African context. This chapter describes the analytical framework used to analyze the objectives of the study. The description starts by presenting the theoretical foundation behind the explanation of factors affecting decentralization process and performance of water resource management and describe the factors affecting decentralization process and performance in its four dimensions: (i) contextual factors and initial conditions, (ii) characteristics of decentralization process, (iii) characteristics of central government/basin-level relationships and capacities and (iv) the internal configuration of basin-level institutional arrangements. Next, taking into account the theory, the chapter introduces the hypothesis to be tested and present the models (decentralization process and decentralization performance models) used to test the specified hypothesis.

Keywords Decentralization process and performance · Hypothesis testing · River basin

3.1 Theoretical Considerations

Based on previous work, we can set several hypotheses with regards to the trajectory of the decentralization process and its performance. We follow the analytical framework suggested by Blomquist et al. (2005, 2010), and Dinar et al. (2007) that uses concepts such as *incentives* for stakeholders to act (e.g., the government to decentralize, the water users and other stakeholders to take on responsibilities),[1] *principal-agent relationships* (referring to the transparency and enforcement possibilities in contractual agreements between the stakeholders to carry out certain functions), *transaction costs* (in terms of time and money to achieve institutional change) as

---

[1]Stakeholders in the basin may include individuals, groups and governments (from local to federal).

© The International Bank for Reconstruction and Development/The World Bank 2016
J. Mutondo et al., *Water Governance Decentralization in Sub-Saharan Africa*,
SpringerBriefs in Water Science and Technology, DOI 10.1007/978-3-319-29422-3_3

well as the level of *influence*, determined *inter alia* by the degree of *information asymmetry* between different actors and social groups in the basin and outside the basin.

In addition to the specific local context of the decentralization process, an important issue to be addressed is what to measure and how to measure. Decentralization of decision-making is not an aim per se. It is recommended because experience over the past decades has shown that when decision-making is centralized and local conditions are not taken appropriately into account, then accountability of decision-makers is weak, and water resources management is inadequate. Thus, it is necessary to develop indicators to (a) define decentralization as a concept, and (b) define and measure changes in water resources management outcomes when the institutional arrangements have changed (Blomquist et al. 2007).

We start with a proposed definition of decentralization, which is based on (a) an increase in transparency in decision-making, and (b) a substantial increase in stakeholder involvement in decision-making, including measures to accord financial self-sufficiency. Acknowledging that each case is different, the baseline used for analysis would be the intention to decentralize as expressed by legislation in a certain country, and by the initial statement of objectives of the respective organization that is being analyzed. The implementation of this intention would then be evaluated by taking into account (a) the existing institutional framework, (b) the process, (c) the political economy, and (d) the results. Decentralization can be seen as a reform process and, as such, can be affected by other processes that take place in parallel. Forces initiating and affecting the decentralization process stem from societal structure in the basin and outside the basin: some of these forces are the initiation of the process, the interests leading to the reform (top-down or bottom-up), and rules governing the initiation and approval of organizational change. These are discussed at length in Blomquist et al. (2005).

Furthermore, the concept of *path dependency* plays a major role in the process of institutional reform (Saleth and Dinar 2004: 264). Path dependency is an important aspect of the decentralization in SSA, due to the nested organizational structures of many river basins that are part of a transboundary river basin organization and possibly international treaties. The process by which decentralization measures are introduced is expected to affect implementation, and thus performance, and therefore needs to be taken into account. The costs and benefits encountered by different stakeholders as well as power relations between them are also considered as important variables in our analytical framework (Saleth and Dinar 2004: Chap. 4).

3.2   Hypotheses: Analysis of Variables

For the purposes of developing the analytical framework, we assume that "management at the lowest appropriate level" usually implies the active involvement of different stakeholders, including users, at various levels related to the river basin. *Appropriate* in this context implies that not all stakeholders need to be involved in all decisions and management activities, but that this is a flexible concept that would be adapted to each river basin, depending on local conditions. It is important to note that increasing stakeholder involvement is not the end of the inquiry, and there are several important related questions. If such active involvement of stakeholders is stable, how can it be translated into effective resource management and high performance level? What factors might we expect to affect the likelihood of stakeholder involvement turning into effective basin-level resource management (as distinct from mere stakeholder consultation, or the collapse of stakeholder involvement)? If stakeholder involvement is translated into basin-level management, how can the active involvement and the effective resource management be sustained over time and changing conditions? What factors might account for the longevity of decentralized arrangements in some cases and their demise in others? Guided by these research questions, we identify four sets of variables under the major headings (contextual factors and initial conditions, characteristics of the decentralization process, characteristics of central government/basin-level relationships and capacities, and the internal configuration of basin-level institutional arrangements) with hypotheses about their impact on the process of decentralization of river basin management and its performance. Those variables and hypotheses incorporate ideas identified in Mody (2004) and Blomquist et al. (2005). They are used here for translating the theory to analytical hypotheses. For each key variable, we develop a list of variables that could capture the expected relationship as follows:

3.2.1 Contextual Factors and Initial Conditions

The literature on decentralized water resource management indicates that the outcome of decentralization is partly a function of the initial conditions that prevail at the time a decentralization initiative is attempted (*path dependency*). These initial conditions are elements of the economic, political, and social context of the decentralization effort. Several variables that could capture such conditions are detailed below.

Level of economic development of the river basin region measures the ability of the basin stakeholders to commit financial and other resources necessary to the decentralization process, in addition to central government provision of support for the decentralization effort. The literature on decentralized water resource management indicates that successful decentralization must include some degree of financial autonomy (Cerniglia 2003; Musgrave 1997). Sustaining this financial autonomy often depends upon the establishment of some form of water pricing or tariffs, having the users obeying such payments, and having the proceeds remain within or return to the basin.

Thus, decentralizing management to the basin level, developing and maintaining the institutional arrangements for basin-level management, and implementing any form of financial autonomy imply that some financial resources at the basin level will have to be committed to the decentralization effort. *This in turn implies that basins that have a level of economic development that can sustain those resource commitments are (all other things being equal) more likely to achieve sustainable success in decentralization.*

Initial distribution of resources among basin stakeholders is an important contextual factor in the development and successful implementation of a decentralization initiative. We also refer to the impact of climate change on the variability of water flows in the basin as a measure of resource availability. This variable has interesting and complex properties, however. On the one hand and more obviously, extreme disparities in resource endowments among basin stakeholders can imperil decentralization success. If some privileged stakeholders may anticipate being worse off, they are unlikely to support the decentralization process and may even try to derail it. And if other stakeholders are so destitute as to be unable to bring any resources of their own to the decentralization initiative, they may rationally elect not to participate, even though more effective resource management would promise to improve their situation in the long run. On the other hand and less obviously, some inequality of initial resource endowments may facilitate action by enabling some stakeholders to bear the costs of taking a leadership role (Blomquist 1988; Ostrom 1990).

Thus, some inequality of resource endowments is not necessarily lethal to a decentralization initiative, and may even facilitate it if better-situated users are willing to lead (Dinar 2009). Extreme inequality, however, may be detrimental or even derail the decentralization effort. The distribution of resource endowments among the basin stakeholders is therefore an important contextual variable affecting the prospects for successful decentralization. *We hypothesize that the relationship between level of inequality of resource endowments and successful decentralization is quadratic, with the greatest positive impact at a certain level of inequality and lower or negative impacts at both lower and higher levels of inequality of resource endowment distribution.*

3.2.2   Characteristics of the Decentralization Process

Certain conditions or characteristics of the decentralization process itself may affect the prospects for successful implementation. Two necessary conditions of a decentralization initiative are (a) a devolution of authority and responsibility from the center, and (b) an acceptance of that authority and responsibility by the local or regional units. Whether (a) and (b) both occur will depend in part upon why and how the decentralization takes place.

Top-down, bottom-up, or mutually desired devolutionare ways of characterizing the decentralization initiative: In some cases, central government officials may have undertaken resource management decentralization initiatives in order to solve their own problems—e.g., to reduce or eliminate the central government's political accountability for past or current resource policy failures, resolve a budgetary crisis by cutting their financial responsibility for selected domestic policy areas

(Simon 2002), respond to pressure from external support agencies to formulate a decentralization initiative as a condition of continued receipt of financial support. In other cases, it is "bottom-up" pressure from the stakeholders that leads to the decentralization (Samad 2005). In still other cases, the decision to decentralize resource management to a lower and more appropriate level may have been the outcome of a process of mutual discussion and agreement between central officials hoping to improve policy outcomes and local stakeholders desiring greater autonomy and/or flexibility. *All other things being equal, we can anticipate that because decentralization initiatives require active basin-level stakeholder involvement, they are more likely to be implemented successfully if undertaken under the latter (bottom-up) circumstances than under the former (top-down).*

Existing local-level governance arrangements contribute to continuation: The literature suggests that decentralization initiatives are more likely to be accompanied by active involvement of basin stakeholders if existing community (village, tribe) governance institutions and practices are recognized and incorporated in the decentralization process. This observation has a transactions costs explanation, too: the costs (primarily in terms of time and effort) to basin stakeholders of relating on familiar organizational forms are expected to be smaller than the costs of relating to an additional set of organizational arrangements. In contrast, decentralization initiatives that feature central government construction of new sets of basin-level organizations that are largely separate from existing and traditional community governance institutions may face higher costs in achieving basin stakeholders' participation, resource commitments, and acceptance of decisions as legitimate. This does not mean that no new institutions will be created in order to achieve basin-scale management. In fact, new institutions will often be needed to promote communication and integrate decision-making across communities within a river basin. *Rather, all other things being equal, decentralization initiatives are more likely to succeed in gaining stakeholder acceptance if they are based upon, and constructed from, traditional community governance institutions and practices (i.e., take account of existing social capital).*

3.2.3 Characteristics of Central Government/Basin-Level Relationships and Capacities

Because successful decentralization requires complementary actions at the central government and local levels, other aspects of the central-local relationship can be expected to affect that success. Accordingly, our study includes a set of political and institutional variables having to do with the respective capacities of the central government and the basin-level stakeholders, and with the relationship between them.

The extent of devolution of responsibilities and decision-making: A decentralization policy initiative announced by a central government may be only symbolic, while the central government retains in practice control over all significant resource management decisions. Worse still, a decentralization policy can represent an abandonment of central government responsibility for resource management without a concomitant establishment of local-level authority. In better situations, the central government transfers degrees of both authority and responsibility for resource management to the stakeholders.

These differences in the extent of actual devolution that occurs can be expected to affect the prospects for successful implementation of the decentralization policy. Symbolic or abandonment policies are at best unlikely to improve resource management, and at worst will undermine stakeholder willingness to commit and sustain the extent of active involvement necessary for successful decentralization. *All other things being equal, we would expect to see greater prospects for success increasing with level of devolution.*

Local-level experience with self-governance and service provision: In any country, the decentralization of water resource management does not occur in a vacuum. The ability of central government officials to strike a balance between supportiveness and intrusiveness, and the capacity of basin-level stakeholders to organize and sustain institutional arrangements, will in part be a function of their experiences with respect to other public services or responsibilities. The ability of central and local participants to perform successfully will depend on the skills and experiences they have developed.

We would expect that water resource management decentralization initiatives are more likely to be implemented successfully for settings in which local participants have experience in governing and managing other resources and/or public services (e.g., land uses, schooling, transportation).

Economic, political and social differences among basin users: In many countries, the distribution of political influence will be a function of economic, religious, or other social and cultural distinctions. But even if it were not for the connection between these characteristics and political influence, the characteristics themselves can affect successful implementation of decentralization initiatives, through their independent effects on stakeholder communication, trust, and extent of experience in interdependent endeavors.

Economic, political, and social distinctions among basin-level stakeholders are likely to affect the implementation of decentralized resource management efforts. *The greater and more contentious these distinctions, all other things being equal, the more difficult it will be to develop and sustain basin-scale institutional arrangements for governing and managing water resources.*

It is important to add that these are empirical, not prescriptive, observations. Central government officials cannot make distinctions among basin-level stakeholders. Nor should central government officials selectively apply decentralization policies only in relatively homogeneous settings.

Adequate time for implementation and adaptation: While it is obvious that longevity of water resource management arrangements may reflect their success, it may be less obvious that their success may depend on their longevity. Time is needed to develop basin-scale institutional arrangements, to experiment with alternatives and engage in some trial-and-error learning. Time is needed for trust-building, so water users begin to accept new arrangements and gradually commit to sustaining them. Time is needed also to translate resource management plans into observable and sustained effects on resource conditions.

The relationship between time and success in water resource management is complicated. On the one hand, we have already said that adaptability is important as water users need to be able to modify institutional arrangements in response to changed conditions. On the other hand, patience is important because a new approach that has not succeeded can simply erode stakeholders' willingness to commit their time and effort to the next reform. *We may observe a curvilinear relationship, in which successful implementation is less likely to be observed among decentralization initiatives that are very young, but is more likely at longer periods, but could taper off if central government and basin-level arrangements have proved insufficiently adaptable over long periods.*

3.2.4 The Internal Configuration of Basin-Level Institutional Arrangements

Successful implementation of decentralized water resource management may also depend on features of the basin-level arrangements created by stakeholders and/or by the central government.

Presence of basin-level governance institutions may be a prerequisite for successful water resource management. Sustained and effective participation of stakeholders presupposes the existence of arrangements by which stakeholders articulate their interests, share information, communicate and bargain, and take collective decisions. Basin-level governance is essential to the ability of water users to operate at multiple levels of action, which is a key to sustained successful resource preservation and efficient use (Ostrom 1990).

Basin-level water resource management (in other words, a decentralized system) is neither achievable nor sustainable without the establishment and maintenance of basin-level governance arrangements. In the case of SSA, we refer also to situations of rivers that are international in nature. Thus having an agreed upon treaty among the various riparians would also fall under this category of sub-basin interests. *Because the existence of governance arrangements is a necessary, not sufficient, condition of successful resource management, we should not expect to find success everywhere we find basin-level governance institutions, but we should expect to find failure everywhere they are absent.*

Recognition of sub-basin communities of interest: The water management issues in the basin are viewed differently by the stakeholders that share the resource in various parts of the basin, based mainly on the physical conditions and spatial situation of each group. For example, downstream users' perspectives on water quality differ from upstreamers. Users with access to groundwater have different views of drought exposure than surface water users. Municipal and industrial water users do not perceive the value of assured water supply reliability in the same fashion that agricultural water users do (Blomquist and Schlager 1999). Thus, while basin-level governance and management arrangements are essential to decentralized water resource management, the ability of sub-basin stakeholders to address sub-basin issues may be as important. In the case of SSA, we refer also to situations of rivers that are international in nature. Thus having an agreed-upon treaty among the various riparians would also fall under this category of sub-basin interests.

Level of participation of various groups in basin-level decision-making arrangements explains the direction and extent of the decentralization process. Of course, transaction costs of the decentralization process increase, as such assurances are institutionalized, since a larger number of stakeholder organizations within the basin will bring greater coordination costs. *All other things being equal, we would expect that successful implementation of basin decentralization has a positive relationship with level of participation of stakeholders in the process.* However, with a diverse and large number of stakeholders, high transaction costs may become a constraint. *Here too, then, a hill-shaped relation of this variable to successful decentralization may be expected, with the absence of sub-basin organizations and large numbers of sub-basin organizations negatively associated with lower success and greater prospects for success in between.*

Information sharing and communication: The importance of information—more particularly, information symmetry—and opportunities for communication to the emergence and maintenance of cooperative decision-making is relatively well understood. In water resource management especially, in which there can be so many indicators of water resource conditions and the performance of management efforts, forums for information sharing are vital to reducing information asymmetries and promoting cooperation.

Since information will not automatically be perceived the same way by all stakeholders, and the implications of information about resource conditions will differ among these groups, it is arguably as important that there also be institutionalized or other regular forums in which basin stakeholders can communicate. *All other things being equal, we expect to find successful decentralized water resource management more likely where information sharing and communication among stakeholders are more apparent.*

Mechanisms for conflict resolution are needed to prevent disagreements from arising. Resource users can and will disagree about how well their interests are being represented and protected, about how well the resource management program is working, and whether it is time for a change, about the distribution of benefits and costs, and manifold other issues.

The success and sustainability of decentralized resource management efforts therefore also depend on the presence of forums for addressing conflicts. *All other things being equal, we would expect successful implementation of decentralized water resource management more likely for settings in which forums for conflict resolution exist.*

3.3 The Models

We apply the framework proposed by Dinar et al. (2007) to analyze river basin decentralization processes and performance. This approach is appropriate here, since it includes various institutional variables and their possible impact on the outcome of the decentralization reform. The approach allows for micro-level analysis, given that it is capable at analyzing a decentralization process and performance at a single river basin level.

The relationship between river basin decentralization process and institutional variables is given as:

$$P = g(C, R, I \mid X) \tag{3.1}$$

where P represents a vector of variables indicating the characteristics of the river basin decentralization process (such as length of decentralization, number of institutions created and dismantled, etc.), C is a vector of variables representing contextual factors and initial conditions involved in the reform process (such as river basin GDP and revenues), R is a vector of variables representing the characteristics of central government/basin-level relationships and capacities (such as the nature of distribution of river basin management responsibilities), I is a vector of variables indicating internal configuration of basin-level institutional arrangements (such as the organizational structure of the basin organization) and X is a vector of other variables associated with the specific river basin (such as river basin size, population etc.).

The relationship between river basin decentralization performance and institutional variables is given as:

$$S = f(C, P, R, I \mid X) \tag{3.2}$$

where S is a vector of river basin decentralization performance indicators and the other variables are defined as described above. The analytical institutional economic framework described above is used to access qualitatively and quantitatively the decentralization process and performance of water resource management.

References

Blomquist, W., Dinar, A., & Kemper, K. (2007). River basin management: Conclusions and implications. In Kemper et al. (Eds.), *Integrated River basin Management Through Decentralization* (pp. 229–238). World Bank and Springer.

Blomquist, W., Dinar, A., & Kemper, K. (2005). Comparison of institutional arrangements for river basin management in eight basins. World Bank Policy Research Working Paper #3636, Washington, DC.

Blomquist, W., Dinar, A., & Kemper, K. (2010). A framework for institutional analysis of decentralization reforms in natural resource management. *Society & Natural Resources, 23*(7), 620–635.

Blomquist, W., & Schlager, E. (1999). Watershed management from the ground up: Political science and the explanation of regional governance arrangements. Paper presented at the annual meeting of the American Political Science Association, Atlanta, Georgia (p. 54). Retrieved 2–5 September, 1999.

Blomquist, W. (1988). Getting out of the commons trap: Variables, process, and results in four groundwater basins. *Social Science Perspectives Journal, 2*(4), 16–44.

Cerniglia, F. (2003). Decentralization in the public sector: Quantitative aspects in federal and unitary countries. *Journal of Policy Modeling, 25,* 749–776.

Dinar, S. (2009). Scarcity and cooperation along international rivers. *Global Environmental Politics, 9*(1), 107–133.

Dinar, A., Kemper, K., Blomquist, W., & Kurukulasuriya, P. (2007). Whitewater: Process and performance of decentralization reform of river basin water resource management. *Journal of Policy Modeling, 29*(6), 851–867.

Mody, J. (2004). Achieving accountability through decentralization: Lessons for integrated river basin management. World bank Policy Research Working Paper 3346. Washington DC: World Bank.

Musgrave, W. (1997). Decentralized mechanisms and institutions for managing water resources: Reflections on experience from Australia. In P. Douglas & Y. Tsur (Eds.), *Decentralization and coordination of water resource management* (pp. 429–447). Boston, MA: Kluwer Academic Publishers.

Ostrom, E. (1990). *Governing the Commons: The Evolution of Institutions for Collective Action,* Cambridge (p. 280).

Saleth, R. M., & Dinar, A. (2004). *The institutional economics of water: A cross-country analysis of institutions and performance.* Cheltenham, UK: Edward Elgar.

Samad, M. (2005). Water institutional reform in Sri Lanka. *Water Policy, 7*(1), 125–140.

Simon, B. (2002). Devolution of Bureau of Reclamation constructed facilities. *Journal of the American Water Resources Association, 38*(5), 1187–1194.

Chapter 4
Case Studies in Mozambique, South Africa, and Zimbabwe

Abstract In this study, the analysis of factors explaining the level of decentralization process and performance of river basin management is run using two main approaches: (i) the case study approach and (ii) an econometric approach, of which the framework is described in Chap. 3. The case study approach was implemented in three countries (Mozambique, South Africa and Zimbabwe) during the first phase of the project and it aimed to test the survey instrument used to collect data in the African countries (phase two) and understand the factors affecting the decentralization process and performance of water resource management in these countries. The results showed that problems like the poor or unequal access to water resources by local stakeholders or the lack of autonomy by the local water agencies seem to be generalized in the studied countries. On the other side, factors such as the presence of basin-level governance institutions and the level of information sharing are likely to be in favor of the decentralization process in the three studied catchments, while the composition of catchment boards were not in favor of decentralization process mainly in Mzingwane catchment in Zimbabwe.

Keywords Decentralization process and performance · River basin · Water laws and policies

4.1 Data Collection and Analysis

In order to analyze the water governance and decentralization process in Sub-Saharan river basins, this study selected a sample of three Southern African catchments: Inkomati (South African part) in South Africa, Limpopo (Mozambican part) in Mozambique, under the responsibility of the ARA-Sul agency, and Mzingwane (the Zimbabwean component of the Limpopo river basin) in Zimbabwe. The choice of the three countries for the case studies is due to the interest of comparing countries where water laws and policies were designed and implemented in the early '90s and therefore are affected by the IWRM paradigm and particularly by the idea of decentralization of water management. Another factor of choice was the geographic

© The International Bank for Reconstruction and Development/The World Bank 2016   27
J. Mutondo et al., *Water Governance Decentralization in Sub-Saharan Africa*,
SpringerBriefs in Water Science and Technology, DOI 10.1007/978-3-319-29422-3_4

proximity of the catchments, which followed the hypothesis that diversity in the decentralization process and performances can exist also in river catchments situated in the same geographic area.

This study used both primary and secondary data. Primary data were collected, in the three studied basins, using a structured questionnaire.[1] For data collection, this study employed a non-random (purposive) sampling, which consists of selecting respondents in a deliberative fashion in order to achieve certain objectives (Prinsloo 2008). For instance, respondents with the best knowledge and experience in the river basin decentralization process were deliberately chosen to answer the questionnaire, since the main objective of the study is to access the impact of institutional factors on river basin decentralization process and its performance. This technique is appropriate for case studies in which a small sample composed of key informants is selected from the target population (Saunders et al. 2007).

The target population of the structured questionnaire was identified as the staff members of the river basin organizations. In Zimbabwe, the structured questionnaire was mainly administered to the Zimbabwe National Water Authority (ZINWA) officials, which lead the Mzingwane catchment council. In South Africa, respondents were officials from the Inkomati Catchment Management Agency (CMA) and the Department of Water Affairs and Forestry. In Mozambique, respondents were officials of the ARA-Sul, water user associations, producers' associations, and government agencies, such as the Chokwe Hydraulic Public Enterprise (HICEP), the Baixo Limpopo Irrigation Scheme (BLIS), and the National Directorate of Water (DNA). Respondents to the structured or to the semi-structured questionnaires either provided factual data or expressed their knowledgeable opinion in terms of performance of the basin decentralization process.

In the addition to the purposive sampling technique, a random sampling was applied and it brought very different samples in the three studied catchments: one structured questionnaire was filled in South Africa and Zimbabwe, and twenty-seven structured questionnaires were filled in Mozambique. In order to collect primary data from a sufficiently larger sample, semi-structured questionnaires[2] capturing information about the decentralization process were administered to 125 randomly selected water users in Zimbabwe. Additionally, twenty non-structured questionnaires were finally administered in the Inkomati WMA in South Africa. A detailed explanation of the collection methods and questionnaires used, as well as the list of interviewed stakeholders in the three case studies is available in Matsinhe et al. (2012), Chiwbwe et al. (2012), Musinake et al. (2012), and Mutondo et al. (2011).

Primary and secondary data collected in the three case studies do not allow a statistical quantitative treatment, and this is due to the limited significance of the

---

[1]The questionnaire is composed of five major sections, namely (1) river basin organization identification, (2) river basin characteristics, (3) decentralization process, (4) decentralization performance, and (5) basin comparisons. The questionnaire used to collect data for the case studies is presented in Mutondo et al. (2011).

[2]The semi-structured questionnaires are presented in Musinake et al. (2012) and Matsinhe et al. (2012).

collected data and to the different weight of the three catchments in the final dataset. This study uses a comparative analysis method, consisting of a qualitative comparison of the collected data with the hypotheses presented in the previous chapter about the impact of selected variables on the decentralization process and performance. The results of this qualitative process are detailed in the Annex A. This approach, therefore, does not estimate the impact of studied variables on the river basin decentralization process. It rather allows describing those variables in the studied river basins and comparing their observed likely impact on the decentralization process with the hypotheses made.

4.2 Historical Context of the Three Studied Countries

Before comparing the results from our analytical framework in the three studied basins, it is important to review and compare the historical political setting in the three countries as they influence the outcome of decentralization process and performance. After the description of the political setting, we compare the development of water laws and policies as they are also important factors in the outcome of the decentralization process and performance. Finally, we compare the results of the analytical framework and draw conclusions for the studied three basins.

While unified, South Africa reached a status of independent state within the British empire in 1902 as a result of the Anglo-Boer War. The instauration of the apartheid regime by the Afrikaners with the support of the British crown kept South Africa away from democracy until the free elections of 1994, which brought Nelson Mandela to power and allowed the formulation, in 1996, of the Constitution. This Constitution (known as final, with reference to the "transitional" one, prepared in 1993 by de Klerk's National Party and the African National Congress to open the door for the democratic elections) included some fundamental rights, among which access to water and sanitation represents a crucial component of modern South Africa's society.

Mozambique was under the Portuguese colonial rule until 1975, when the Salazar fascist regime was forced to abandon the country by the *Frente de Libertação de Moçambique* (FRELIMO), headed by Eduardo Mondlane, who died in 1969, and Samora Machel, who became the first president of independent Mozambique. Shortly after independence, in 1981 a devastating civil war exploded between FRELIMO and the *Resistência Nacional de Moçambique* (RENAMO), a movement created in the late '70s as a "guerrilla force" by Ian Smith's Rhodesia to contrast the Mozambique government, which supported United Nations sanctions against the racist rule in Southern Rhodesia. Passed under the support of South Africa after 1980, year of the fall of Smith's regime, RENAMO grew rapidly and became a powerful challenger to FRELIMO, which responded militarily, starting a bloody and destructive domestic conflict that finally ended in 1992 (agreements of Rome) when democratic elections under the supervision of the United Nations were prepared for 1994 and determined a FRELIMO convincing victory.

Zimbabwe's independence dates back to 1980, when Canaan Banana and Robert Mugabe, leaders of the national resistance movement ZANU were elected president and prime minister, respectively. Historically, the former Southern Rhodesia, so called after the South African businessman and politician Cecil Rhodes was part of Zambezia, a territory including today's Zambia (north-eastern Rhodesia) and Zimbabwe. C. Rhodes, through its British South African company, established since 1890 treaties with local populations and obtained concessions to exploit natural resources in the whole Zambezia region and Nyasaland (today's Malawi). During sixty years, the white British minority (about 2,50,000 people at its apex) ruled over about ten million Africans in Southern Rhodesia, taking advantage also of the internal conflicts between the two main ethnic groups of natives: the Shona and the Ndebele. In 1963, after the independence of Zambia and Malawi, Ian Smith, prime minister of Southern Rhodesia, also declared unilaterally the independence of the "Republic of Rhodesia." The United Nations did not recognize the state and put sanctions against the racist regime of Smith. Starting from the '50s, independence movements ZANU (Shona) and ZAPU (Ndebele) started to mobilize in the country and, by the end of the '60s, a real war exploded against the regime of Ian Smith, who was defeated in 1979. Under the United Kingdom supervision, a transitional government prepared the first national free elections, which took place in 1980 and gave the power to the ZANU party.

For the purposes of this work, we can consider the dates corresponding to the advent of a democratic rule in the three studied countries are respectively 1980 (first democratic elections) for Zimbabwe, 1992 (end of the civil war) for Mozambique, and 1994 (first democratic elections) for South Africa. In Mozambique the civil war exploded just after the independence (1975) delayed the normalization of the democratic rule until 1994.

4.3   Water Laws and Policies

During the colonial era, water resources were regulated in the three studied countries by the Portuguese legal framework in Mozambique, and by the English and Roman-Dutch framework in South Africa (until the end of apartheid) and Zimbabwe. Following these legal systems, in South Africa and Zimbabwe water resources were regulated using the riparian principle, which states that landowners bordering a water body (riparian owners) were entitled to make reasonable use of the water (Musinake et al. 2012). Water rights (private) were allocated in perpetuity on the basis of land holding. In Mozambique, private property rights on water were also admitted until the end of the colonial rule.

The advent of a democratic political system in the three countries introduced new constitutional rights and rules, among which water had a prominent role. The new national water acts date back to 1991 in Mozambique (followed by the national water policy in 1995), and to 1998 in South Africa (followed by the national water resource strategy in 2004) and Zimbabwe (Musinake et al. 2012; Chibwe et al. 2012; Matsinhe et al. 2012).

The mid-nineties were the years of the global dissemination of the integrated water resource management (IWRM) principles, expressed clearly during the 1992 Dublin International Conference on Water and Environment. Among the IWRM principles, stakeholders' participation is the one calling for river basin management at the lowest appropriated level. This refers to the idea of decentralization of water policies implementation. Blomquist et al. (2005) indicate that effective decentralization requires devolution of authority and responsibility from the center, and acceptance of that authority and responsibility by local entities in the basin. In other words, following the subsidiary principle, the design and implementation of water management and allocation policies are transferred from the state to local institutions.

The legislators in the three studied countries followed the IWRM and decentralization principles in the preparation of the national laws and policies. In Mozambique, the first national water law created five regional water administration agencies (ARAs). These ARAs were created in order to implement integrated water resource management at the river basin level across the country. The five ARAs are responsible for the management of the thirteen river basins in the country. In Zimbabwe, the new water act established catchment and sub-catchment councils to manage seven major watersheds identified in the country in an attempt to decentralize management of water. In South Africa, the 1998 Water Act established nineteen water management areas (WMAs). Within each WMA, the law established the progressive creation of catchment management agencies (CMAs), sub-catchments entities (Catchment Management Committees—CMCs) and water user associations (WUAs).

4.4   Comparative Analysis of the Basin Case Studies

While the detailed results presented in Annex A originate from a limited and not uniform sample, and the data collection and processing suffered from gaps and missing information, the collected material in the three studied water management areas is rich and allows a comparison among the different cases.

For the reasons mentioned above, this comparison can only have a pure descriptive value, not being based on sufficiently sound statistical ground. The following exercise is therefore an attempt to summarize and sort the information collected and described in Annex A by comparing the observed results of the survey conducted in the three countries with the hypotheses made in the literature (Dinar et al. 2007) about the possible impact of the analyzed factors on the decentralization process and performance.

In Table E.1, the four groups of variables included in the analytical framework are presented, and their possible impact on the decentralization process of the three studied water management areas is indicated. The positive, negative, or contrasted impact that can be assumed for each variable is the result of the observed situation in the field, compared with the hypotheses made by Dinar et al. (2007).

In terms of **contextual factors and initial conditions**, the *level of economic development* in the country and in the catchment at the moment when decentralization started was very low in Zimbabwe (degradation of the economic system) and Mozambique (aftermath of the civil war), while a growing economy and an increasing interest from external donors made the situation in South Africa better off. The studied variables indicated for all three basins critical situations regarding the *distribution of water resources among local stakeholders*. Distribution of access was indicated as very skewed in South Africa and Zimbabwe, and generally resulting in very poor access in Mozambique. Authors of the case studies interpreted this situation as potentially negative for the decentralization process and performance, as inequalities and poor initial endowment were seen as a factor of exclusion of disadvantaged stakeholders from the process. The *level of managerial skills* by local stakeholders was seen as sufficient in Zimbabwe, while in South Africa it was only developed after the implementation of the ICMA, and in Mozambique it is low.

With regard to the **characteristics of the decentralization process**, the *type of devolution of the decentralization process* was seen as very top-down in Zimbabwe and Mozambique, where the process is mainly a shift of the state power to state agencies (ZIMWA and ARA-Sul) depending on the respective ministries, and more mutually desired process in South Africa, where *efforts to involve local stakeholders* and make them part of the process from the beginning are more evident. The efforts in South Africa result in a more diversified *composition of the catchment boards* and in a more active participation by local representatives. Particularly evident in Zimbabwe was the gender issue represented by a limited access to managerial position by women. This situation was not reported at the same level of importance in the remaining two case studies.

In terms of **central government/basin-level relationships and capacities**, the studied factors indicate that the devolution of power (particularly at the financial level) is still relatively low in the three observed catchments. The *source of the river basin budget* is the state in South Africa (no data available for Mozambique); while in Zimbabwe it comes from stakeholders' tariff payments. The first two cases show a lack of financial autonomy by the river basin organization, while in the case of Zimbabwe the majority of river basin resources are from river basin stakeholders which might guarantee financial sustainability over time. However, the low contribution from the government might indicate a lack of government commitment in the decentralization initiative. An important *share of the water tariffs collected from the local users* remain in the basin in Zimbabwe, where 75 % of the collected tariffs stay locally, but only 1 % go to stakeholders institutions, while the remaining part is for ZINWA. Conversely, in South Africa and Mozambique none of the collected water tariffs remain in the basin. The *level of management authority given to basin stakeholders* is still very low in Mozambique, while no data was available for South Africa. In Zimbabwe the establishment of the Mzingwane catchment council and the abolishment of the water right system for the renewable water permits allocated locally were seen as a step toward devolution of management authority to locals.

Finally, in terms of **configuration of basin-level institutional arrangements**, the *presence of basin-level governance institutions* and a well-structured hierarchy of managing organizations can be seen as a potentially positive factor for water decentralization process and performance in the three case studies. *Information sharing* is a key factor for the success of decentralization processes, as it reduces asymmetries among stakeholders and fosters cooperation. In the three studied catchments, efforts to establish forums and supports for information dissemination and sharing have been observed. In Zimbabwe, the use of English and Western protocols and practices during council meetings marginalize disadvantaged community representatives. Similar results have been observed in South Africa, where disadvantaged stakeholders do not participate actively in council meetings. *Forums for conflict resolutions* exist in the Inkomati (water tribunals) and in the Mzingwane, but according to case study authors, only the water tribunals have effectively been active to solve conflicts. In the Mozambican portion of the Limpopo, the basin committee works like a forum to hear disputes, but without authority to solve them.

Table E.2 contrasts the interpretation by the respondents of the **decentralization performance** in the three catchments. As for the previous results, these comparisons must be taken with all precautions as they come from individual perceptions from a limited sample of interviewees.

In terms of the *level of accomplishment of river basin objectives*, in the Mzingwane basin respondents indicated that while water conflict problems were mostly solved, water allocation still remains a main issue. Similarly, in the Inkomati catchment the objectives to reduce water conflicts, water scarcity, and to improve water quality were partially reached. In Mozambique, while primary data for this aspect were not available, the case study's author pointed out that the main catchment objectives are still far from being reached. In terms of the state of issues related to river basin stressed resources after decentralization, while in Zimbabwe some problems (water scarcity and water conflicts) were considered less important after decentralization, other problems (river ecology and land degradation) worsened after the decentralization process. In South Africa, with the exception of water availability and water conflicts, both reduced, all other problems are considered at the same level of acuity as before the process started. In Mozambique, respondents consider that the severe conditions of water resources before decentralization did not improve after the process was implemented. Finally, respondents from all catchments consider as a positive performance the introduction of *renewable water permits* allocated by the local authority in substitution of the permanent water rights that prevailed before the decentralization process.

4.5   Limitations of the Case Studies

The comparison of data collected from the three catchments studied allowed the formulation of interesting hypotheses on the possible impact that the observed factors can have on both decentralization processes and performances. The interviewees' points of view made it also possible to compare their visions in terms of real performance in the three catchments.

The results presented must be considered with the highest precaution for the methodological caveats indicated above, and for the limited sample of the survey. The outcomes of the three case studies were verified, and the hypotheses produced were tested during the second phase of the project, when a continent-wide survey was conducted in Africa using the same structured questionnaire, but adapted to the African context on the basis of the experiences in the three case studies. The continent-wide survey is described in the following two chapters. The detail on major basins and basin organizations in Sub-Saharan Africa (SSA) countries is presented in Annex B. The data collected from the survey were processed using econometric models, based on the same analytical framework mobilized for these case studies. The results we got from the three case studies are compared with the African survey and conclusions are drawn in Chap. 7.

References

Blomquist, W., Dinar, A., & Kemper, K. (2005). Comparison of institutional arrangements for river basin management in eight basins. World Bank Policy Research Working Paper #3636, Washington, DC.

Chibwe, T., Bourblanc, M., Kirsten, J., Mutondo, J., Farolfi, S., & Dinar, A. (2012). Reform process and performance analysis in water governance and management: A case of study of Inkomati Water Management Area in South Africa. Working paper.

Dinar, A., Kemper, K., Blomquist, W., & Kurukulasuriya, P. (2007). Whitewater: Process and performance of decentralization reform of river basin water resource management. *Journal of Policy Modeling, 29*(6), 851–867.

Matsinhe, M., Mungatana, E., Mutondo, J., Farolfi, S., Dinar, A., & Tostão, E. (2012). Reform process and performance analysis in water governance and management: A case of study of Limpopo river basin in Mozambique. Working paper.

Musinakle, G., Dzingirai, V., Mutondo, J., Farolfi, S., & Dinar, A. (2012). Reform process and performance analysis in water governance and management: A case study of Mzingwane catchment in Zimbabwe. Working paper.

Mutondo, J., Farolfi, S., Dinar, A., & Hassan, R. (2011). Water governance and decentralization in Africa: A Framework for reform process and performance analysis. An application to three southern African river basins. Working paper.

Prinsloo, A. (2008). *A critical analysis of the LARD sub programme in Gauteng Province of South Africa.* Unpublished MInst Agrar Dissertation, University of Pretoria. Retrieved January 15, 2011, from http://upetd.up.ac.za/thesis/available/etd-08112009-114937.

Saunders, M., Lewis, P., & Thornhill, A. (2007). *Research methods for business students* (4th/5th ed). Harlow, Essex: Pearson.

Chapter 5
Quantitative Analysis: Empirical Models and Data Collection Process

Abstract As mentioned in previous chapters, the analysis of factors affecting decentralization process was done using both qualitative and quantitative approaches. Chapter 3 established the theoretical foundations of the quantitative approach, the hypothesis to be tested and the model used to test the hypothesis in its three dimensions: (i) contextual factors and initial conditions, (ii) characteristics of decentralization process and (iii) characteristics of central government/basin-level relationships and capacities. However, it did not describe the empirical model used to analyze the factors affecting decentralization process and performance. This is the objective of this chapter, where the variables used in the three dimensions listed above and the procedures used to construct these variables are described. Furthermore, the present chapter highlights the process used to collect the data, the challenges faced during data collection and mechanisms used to mitigate these challenges and assure data quality.

Keywords Data collection · Decentralization process · Performance · River basin

5.1 The Empirical Models

We apply empirically the analytical institutional economic framework described in Chap. 3 and presented through Eqs. 3.1 and 3.2. The empirical approach taken in this chapter builds on and extends the framework used by Dinar et al. (2007) to address new developments and experiences in SSA basins, to introduce situations common in SSA basins, and to account for likely climate change impacts believed to affect decentralization considerations and performance in SSA.

As indicated in Eqs. 5.1 and 5.2 below, we postulate that the characteristics of the decentralization process (P)[1] and the level of the decentralization success/progress (S) can be estimated using a set of variables that include: contextual factors and initial conditions; characteristics of central government/basin-level relationships

---

[1]Variables represented by a bold letter indicate a vector.

© The International Bank for Reconstruction and Development/The World Bank 2016
J. Mutondo et al., *Water Governance Decentralization in Sub-Saharan Africa*,
SpringerBriefs in Water Science and Technology, DOI 10.1007/978-3-319-29422-3_5

and capacities; internal configuration of basin-level institutional arrangements; and a set of "other" variables, identified as necessary. These groups of variables and their relationships were discussed in Chap. 3 and in Blomquist et al. (2010), and Dinar et al. (2007), and will be used in this chapter as well. In addition, we use two new variables that have not been explicitly used in Dinar et al. (2007). One variable indicates whether or not the basin in question is governed by an international river basin organization, under an international treaty. International river basin organizations may include many tributary basins, all constitute the international basin. The second variable measures the likely impact of climate change on precipitation or runoff in the river basin.

The first equation (Eq. 5.1 below) explains a certain phenomenon in the basin, such as specifics of the decentralization process, measured by the levels of P. The second equation (Eq. 5.2 below) explains the level of success/progress of the decentralization process, measured by S.

The set of equations used in the estimation of the first relationship takes the following shape:

$$P = g(C, R, I \mid V, B, X) \tag{5.1}$$

where:

P is a vector of characteristics of the decentralization process;
C is a vector of contextual factors and initial conditions;
R is a vector of characteristics of central government/basin-level relationships and capacities;
I is a vector of internal configuration of basin-level institutional arrangements;
V represents the climatic conditions (precipitation or runoff) in the basin;
B is a dichotomous variable indicating whether or not the basin is governed under an international river basin treaty/organization; and
X is a vector of 'other' variables, identified as necessary.

A general relationship for decentralization success/progress, is given as follows:

$$S = f(C, P, R, I \mid V, B, X) \tag{5.2}$$

where S is a vector of performance indicators of the decentralization in the river basin. All other variables are as defined earlier.

We have several measures of success and several measures of the decentralization process. One possible way to measure success is by using a dichotomous variable that takes the value 1 when decentralization was initiated and 0 when no decentralization took place, in spite of government intent. A second way of describing success is to measure normatively the extent of achieving several important original goals of the decentralization reform. The success variable was computed as an aggregation of the success ratings over the different reported decentralization objectives because the KMO-statistic[2] of some individual success objective variables was very low. A third way of measuring progress of decentralization is by comparing performance between present and the pre-decentralization period. Performance variables may include: level of participation, local responsibility, financial performance, and economic activity. By comparing before and after values, we are just comparing change levels of each of the variables included in the comparison of before and after decentralization.

The first specification explains whether or not a decentralization process was initiated (Eq. 5.1). We expect that it takes some level of the contextual factors (C), as well as characteristics of the central government/basin-level relationships and capacities (R) to initiate the decentralization. However, we are not sure about the direction of the impact of various internal configurations of basin-level institutional arrangements (I). Some existing water user associations may work in opposite directions. We expect that harsh climatic conditions (V) will be associated with higher likelihood of establishing river basin organizations and existing international treaties or international river basin organizations (B) that overrules the basin will help also in establishing the domestic RBO. We actually had to use the linear probability model (LPM) approach because of the small number of observations. LPM is not bounded between zero and one, and captures the intensity of the relationship between the dependent and the independent variables.

Several variables could help shed light on the decentralization process. Few are probably of special interest, as they contrast observations across river basin decentralization processes under a variety of situations.[3] The length of the decentralization process, *YrsDecentralization*, the transaction costs of the process, measured by several variables such as *Institutional Dismantled*, *PoliticalCost*, and the level of involvement of the stakeholders, *WUA Involvement*, are a few that caught our attention. Estimation procedures explaining *Intuitional Dismantled*, *Political Cost*, and *YrsDecentralization* use an OLS procedure as values of these variables are dummies or continuous. Table E.3 summarizes the various equations we specified for relationship 1 (Eq. 5.1), and the hypothesized directions of impact of the independent variables, based on the theory developed in Chap. 3.

We identified several variables that serve to measure decentralization success or progress. The estimates of relationships using the first two approaches (that have been mentioned earlier) to measuring success/progress imply LPM, TOBIT and OLS estimation procedures. We use the variable *Success over Objective* (calculated as an aggregation of the success over all objectives) to reflect achievement of various goals the decentralization process was aimed to achieve. We applied LPM, TOBIT and OLS procedure to estimate that relationship as well. Because we are not sure that the values measured are distributed normally, we cannot use GLM as it may provide

---

[2]Kaiser-Meyer-Olkin (KMO) statistic, predicts if data are likely to factor well, based on correlation and partial correlation. The KMO overall statistic is used to decide whether or not to include a variable in the PC analysis. KMO overall should be 0.60 or higher to proceed with factor analysis. Variables with KMO statistic lower than 0.60 should be dropped from the PC analysis.
[3]For definition of the variables see Annex C.

biased estimates. Thus we use the TOBIT procedure that assumes a Poisson distribution.

Finally, we construct the additional variable, *Problems After*, to explain the performance of the decentralization process. *Problems Before* and *Problems After* are 2 variables for which we did use Principal Component. Table E.4 summarizes the estimation procedures of the various equations we specified for estimating relationship 2 (Eq. 5.2), and the hypothesized directions of impact, based on the theory developed in Chap. 3.

5.2   Data Collection Process

A survey instrument in Dinar et al. (2005) was modified to collect the data needed for estimating the model equations in Sub-Saharan Africa described above. It was first pre-tested on three river basin organizations (RBOs)[4] prior to being modified, translated from English to French and Portuguese, and sent to the identified offices of the river basin organizations in the various states. The English version of the survey instrument is presented in Annex D. A total of twenty-seven RBOs in SSA known to have undergone decentralization to various extents are included in the final dataset we analyze.

Data collection was undertaken by PEGASYS, a consulting firm in South Africa with widely established contacts with water sector agencies in SSA countries. Data collection was completed after several iterative processes of data entry and quality assurance reviews by the authors. Additional rudimentary statistical tests were undertaken to identify, verify, and correct outliers in the dataset. The questionnaires were filled by staff from the basin organizations. All questions, especially those related to performance of the decentralization reform, required objective rather than subjective answers. We intentionally approached local authorities following the reasoning suggested by Alderman (2002), who observed that local authorities appear to have access to information that is not easily captured in official census datasets.

5.2.1   The Potential Final Set of Basins Included in the Study

The basis for the identification of the potential river basin organizations (RBOs) in SSA was ANBO, AMCOW and GTZ (2012), which identifies ninety-nine basins in Eastern, Western, Southern, and Central Africa (Table E.5).

This list of basins (Table E.6) was assessed by PEGASYS and revised, based on a set of investigation approaches such as establishing contacts with local NGOs,

---

[4]The river basins where the questionnaire was tested are Inkomoati in South Africa, Limpopo in Mozambique, and Mzingwane in Zimbabwe.

regional agencies, and known water projects. This process yielded a much more detailed list of 121 basins and their decentralization status (Table E.7).

As can be seen from Table E.7, of the 121 basins, twenty-nine have not started any decentralization activity, and the status of decentralization in twenty-six other basins was impossible to verify. This left us with sixty-six basins that went through decentralization or that have not yet completed the decentralization process.

The final sample was composed of twenty-seven RBOs located in six countries distributed over two of the four SSA regions (four RBOs in two Eastern Africa region countries, and twenty-three RBOs in four Southern Africa region countries). Since the other two regions in the continent, Central Africa and West Africa, do not have decentralization experiences or information about it, the respective basin organizations were not included in the sample. Our sample is quite representative and balanced, representing nearly 30 % of the fourteen Eastern basins, and 44 % of the twenty-three Southern basins that underwent decentralization. It also suggests that we obtained a 41 % response rate. While this response rate is considered barely acceptable in any other place on earth, it is quite significant in SSA.[5] A description of the twenty-seven basins, the country they belong to, and their status of decentralization are presented in Table E.8. The list of the twenty-seven RBOs, including their geographical location, can be found in Fig. F.1 and Table E.9.

5.2.2   The Administration of the Questionnaires

It is the set of these sixty-six basins to whom questionnaires were distributed. The strategy for eliciting responses included: introductory emails followed up by phone calls to identify a focal person; shipment of the questionnaire by email; follow-up on progress by email, as well as phone; clarification sessions with some respondents about difficult questions; review of the received questionnaires and follow-up on particular responses as needed; and translation of the questionnaire into an electronic dataset in Excel. The data collection work was planned for six months (March 2012–September 2012), but actually lasted much longer (March 2012–September 2013) due to communication difficulties that PEGASYS encountered with the respondents.

5.2.3   Quality Assurance Procedures

The electronic dataset was shared with the researchers as it was established over time. There were an overall five rounds of feedback from the research team to PEGASYS. Feedback included inconsistencies in recording missing values (99,999)

---

[5]Another measure of response rate could be obtained from the ratio of questionnaires that were returned, to questionnaires that were sent to potential responding RBOs. Sixty-six questionnaires were sent and twenty-seven were filled, which makes the response rate at 41 %.

and 0 values, replacement of string values with numerical values, and correction of some basic physical information of the basin. Once these inaccuracies had been addressed, the dataset was considered complete, even though some variables were not filled.

In order to increase the response rate, a follow-up survey was sent to the respondents if they did not respond to the survey within a month, and then continued by a telephone follow-up, if necessary. To ensure the highest possible quality, the research team constituted an iterative process of data acquisition and quality assurance reviews. The process involved the compilation of qualitative and quantitative data from a questionnaire, which the agency that collects the data, PEGASYS, distributed.

All responses were checked both by PEGASYS and a graduate student at the University of California, Riverside (UCR), under the supervision of the principal researchers, for errors, which could be critical to the study, such as missing answers to questions, which respondents for one reason or another did not or could not answer. In addition to such a check, a further rudimentary statistical test was conducted on most variables, to identify outliers within the given response range, and to ensure that values are justified. In all cases, the seemingly errors were brought to the attention of the respondents and, in the case of actual errors and/or mistakes, efforts were made toward correction.

5.2.4   Variable Construction

Our questionnaire consisted of fifty-six primary questions, and 245 primary variables (see Annexes C and D). Some of the variables in our data set are naturally correlated to each other. We conducted several principal component (PC) analyses in order to capture the information in these variables and to prevent possible multicollinearity, by combining a set of primary variables into one inclusive PC variable in our estimated relationships. Unfortunately, due to the quality of some of the variables in the dataset, the PC analysis did not yield meaningful results, and could not be used in our analysis (see footnote 3 above). We also used several primary variables to create indices to reflect values that are better expressed on a relative rather than on an absolute scale, or to create dummies that captured key aspects of the decentralization process.

References

Alderman, H. (2002). Do local officials know something we don't? Decentralization of targeted transfers in Albania. *Journal of Public Economics, 83*, 375–404.

Blomquist, W., Dinar, A., & Kemper, K. (2010). A framework for institutional analysis of decentralization reforms in natural resource management. *Society & Natural Resources, 23*(7), 620–635.

Dinar, A., Kemper, K., Blomquist, W., Diez, M., Sine, G., & Fru, W. (2005). Decentralisation of river basin management: A global analysis. The World Bank (Policy Research Working Paper 3637). Washington: The World Bank.

Dinar, A., Kemper, K., Blomquist, W., & Kurukulasuriya, P. (2007). Whitewater: Process and performance of decentralization reform of river basin water resource management. *Journal of Policy Modeling, 29*(6), 851–867.

Chapter 6
Results of Quantitative Analysis

Abstract As mentioned in the previous chapter, the quantitative analysis was performed using data collected in twenty-seven RBOs. This chapter presents the results of statistical analysis. They are split into two subsections: the descriptive statistics and inference of the hypotheses described in analytical and empirical frameworks. The results show that a grass root initiative without government support is not enough to implement sustainable decentralization process as the majority of the basin local institutions and basin stakeholders do not have financial resources and skills, respectively. Therefore, training water user associations revealed to be important for high efficacy of the decentralization process. Additionally, the results show that having water scarcity problems, experiencing longer periods of implementation and having appropriate budgetary support are important drivers of the decentralization process.

Keywords Decentralization process and performance · River basin · Sub-Saharan Africa

6.1 Descriptive Statistics

While we based our entire analysis following the structure suggested in Dinar et al. (2007), due to the reasons indicated in Chap. 5, we had to revise the measurement of some of the variables, and to eliminate several other variables that were not reported due to difficulties of the respondents in SSA basins to assign values to them. This shrunk the usable variables, and reduced the overall number of observations that we could include in the various estimated models. A detailed definition of the variables in our dataset can be found in Annex C (for the variables we created for this analysis). The descriptive statistics of the variables that were included in the analysis is presented in Table E.10.

Table E.10 demonstrates the problems in filling out the questionnaire as the number of variables with full coverage of the entire set of observations fluctuates between ten and twenty-seven. Of the available information, some of the descriptive statistics indicates that about 40 % of the basins were created with a bottom-up approach. In 80 % of the basins that started the decentralization process, RBOs were created. In 58 % of the basins, at least one institution was dismantled during the decentralization process. It is also clear that disputes over water scarcity seem to be more relevant than disputes over allocation. The decentralization process, on average, is about one decade old, ranging between 2 to 30 years. Decentralization processes in SSA started as early as 1979 and as late as 2009 (according to our sample). Finally, climate change may be impacting 76 % of the basins through flow variation, and 68 % of the basins in our sample are part of a transboundary river, governed by international treaty.

6.2 Inference of Our Hypotheses

Following Dinar et al. (2007), we inferred our hypotheses regarding process and performance of the decentralization reform in SSA. Given the few countries in our database, we could not include state-level variables, such as wealth, regime, etc. In addition, we lost several observations due to missing values of some of the variables involved.

6.2.1 Performance of Decentralization (Before and After)

We start by comparing several water management responsibility indicator items before and after the decentralization, using a two-tailed *t-test*. The results of the analyses of four activities (water administration, infrastructure financing, water quality enforcement, and setting water quality standards) are presented in Table E.11.

As can be seen in Table E.11, more water management activities at higher decentralized levels have been reported after the decentralization process, compared with the situation before the decentralization. With ranking of water activities varying between 1 and 5 (with 1 indicating centralized, and 5 indicating most decentralized activity), one can see that there was a significant move of responsibilities toward basin level, and a significant reduction of responsibility at the central government (increase in local responsibility was not significant, and the same is true for increase in state responsibility). A significant increase of responsibilities toward basin level was also reported in the case of infrastructure financing (increase in responsibility at the local level, and decrease in responsibility in the state and central government levels were not significant). A significant increase in responsibility for water quality enforcement at the basin level was reported (insignificant increase in local responsibility, and insignificant decrease in state and central government responsibilities were also reported). A significant increase in responsibility at the basin level was reported for setting water quality standards (no significant changes have been reported for local, state, and central government). As a whole, our sample RBOs have moved after the decentralization process toward more responsibility at the basin level for all four water management decision-making activities. At the same time, these RBOs show a reduction in the central government responsibility in only water administration and water quality enforcement activities. Compared with Dinar et al. (2007), we introduced in this analysis a category of local responsibility (mainly due to the very large size of the basins in SSA, compared to many of the basins in the study by Dinar et al. (2007)). However, by 2013, there is still no progress toward increased responsibilities to the local communities, which suggests difficulty in implementing decentralization toward local actors.

We were also able to get assessments of the severity levels of several issues the basin have been facing, and compare the situation before and after the decentralization. The scales used were: (i) Ranking of severity before decentralization, 0: No problem, 1: Some problem, 2: Severe problem; (ii) Ranking of severity after decentralization, −1: Situation worsen, 0: situation the same, 1: Situation improved. Means of these assessments for each problem item are presented in Table E.12.

Table E.12 suggests that before decentralization, except for floods (with mean value of 0.95), all of the other issues were in the range of "some problem" to a "severe problem." Water conflicts and development issues exhibit the highest level of severity in the sample basins. After decentralization, all six issues have been either stable or improving, with floods, land degradation, and development issues being closer to 1, indicating that the situation related to these issues tended to improve on average. The situation remains on average the same for water scarcity, environmental problems, and water conflicts.

6.2.2 Determinants of the Decentralization Process

We used three decentralization process variables that allowed us to use most of the observations in the dataset. The results of the estimated equations are presented in Table E.13.

The results in Table E.13 indicate very significantly that regardless of the inclusion of the international treaty and the flow variation over time, all contextual factors included, as well as the variables that measure the internal configuration of basin-level institutional arrangements, were significant and follow the expected sign, except the *Creation Bottom-Up* variable. The coefficient of the *Political Cost* is positive and highly significant, suggesting that a higher political cost increases the water users involvement, and may lead to the creation of an RBO as a way to establish the new framework for a cooperative use of the resources. The negative sign on the coefficient on *Creation Bottom-Up*, while opposite to our initial expectations and previous findings (Dinar et al. 2007) is in line with the anecdotal information provided in Chap. 1, and in Mutondo et al. (2011), suggesting that the WUAs that have been established in the RBOs were not well prepared to take off the decentralization process, lacking organizational, legal, and technical skills. This result may indicate that some central government involvement is still needed in SSA basins as a way to transfer not only responsibilities, but also skills to manage the resources under the decentralized arrangement. This support of the central government is needed so that the WUAs creation and implementation process is not "manipulated" by dominant groups and, therefore, is neither equitable nor sustainable. More generally, this finding suggests that *Creation Bottom-Up* is a necessary but not sufficient condition for institutional decentralization.

Being under an *International Treaty* improves cooperation and raises the likelihood of an RBO being created and institutions dismantled. At this point, it may seem that an international treaty that coordinates the various parts of the basin located in different countries may serve as a roadmap for a more effective decentralization and a support tool for users to take the reins of the water resources management in a more stable and accountable setting.

The variable *Disputes over Allocation* has negative and significant coefficient in the equation explaining *WUA Involvement,* and a positive and significant coefficient in the equation explaining *RBO Created.* These results follow our expectations. They suggest that not having sufficient dispute resolution mechanisms lead on the one hand to disengagement of WUAs and, on the other hand, it does provide impetus to the creation of the RBO. Indeed having water conflicts before the decentralization was indicated as the most severe problem (Table E.12).

Results for several water scarcity variables are worth mentioning. *Relative Water Scarcity, Share of Surface Water*, and *Water Flow Fluctuates* are significant and have positive signs. This suggests that water scarcity in the range observed in our sample leads toward more involvement of the WUAs, more likelihood of creation of the RBO, and dismantling of existing institutions in the process of decentralization.

6.2.3 The Decentralization Performance

We were somehow limited in our ability to use the data on all variables that are expected to measure and explain decentralization performance. We remained with only two variables that measure performance, *Success over Objectives* and *Problems After Decentralization.* The results of our regression analyses are presented in Table E.14.

Scrutiny of the results suggests that in spite of having a small number of observations, our model is of high explanatory level and significance. All coefficients are significant and with the expected sign, except for *Water Flow Fluctuates* and *International Treaty*, which are not significant. Adjusted R-squared ranges between 0.964 and 0.998 and F-test values are significant at 1 % and less. The results indicate that higher *Share of Surface Water*, as well as a longer experience with the decentralization process (*Years Decentralization*) enhances the success over the basin's objectives. Lower levels of water scarcity, up to a point, may allow for an easier cooperation and coordination of the users and for a faster accommodation of the decentralization arrangements. In other words, the absence of an acute problem around water availability facilitates conditions for coordination and common approach toward basin solutions. A longer decentralization process may indicate the possibility of the establishment and learning of a cooperative behavior, and the stability of the mechanisms to solve disputes, which translate into a higher social capital accumulation. Contrary to the previous table, the political cost is highly significant and of a negative sign. It could be entirely possible that, as for sharing the benefits of the decentralization process, an excessive level of political costs (through the changes of institutions or the imposition of new duties) may offset any possible short-term gain. Also, it is not because RBOs are created that problems are solved.

Not like in the equations estimating the decentralization process characteristics, here, *Creation Bottom-Up* has a positive impact on the performance of the decentralization. That a higher-level *Governing Body* fosters the accomplishment of the objectives may be an indication of the need of the higher government levels to be active and supportive during the decentralization process. Having a higher *Budget per Capita* is an important factor in having fewer *Problems after Decentralization*, which is an important finding with policy implications. Some other coefficients deserve additional discussion as their coefficients are different in the decentralization process equation (Table E.13) and in the decentralization performance equations (Table E.14), which was expected, based on our theoretical framework (Tables E.3 and E.4). *Political Cost* has a positive sign in the process equations and a negative sign in the performance equation; *Creation Bottom-Up* has a (surprising, but justifiable) negative sign in the process equation, and a positive sign in the performance equation; and *Years Decentralization* has a negative sign in the process equation, and a positive sign in the performance equation.

References

Dinar, A., Kemper, K., Blomquist, W., & Kurukulasuriya, P. (2007). Whitewater: Process and performance of decentralization reform of river basin water resource management. *Journal of Policy Modeling, 29*(6), 851–867.

Mutondo, J., Farolfi, S., Dinar, A., & Hassan, R. (2011). *Water governance and decentralization in Africa: A Framework for reform process and performance analysis*. An application to three southern African river basins. Working paper.

Chapter 7
Conclusions and Policy Implications

The process of water management decentralization in African countries is seen as a means of implementing river basin management at the lowest appropriate level. However, very different stages of implementing decentralization have been observed in practice. This called for a research aiming in understanding the following questions: (i) why do some water agencies succeed more than others? (ii) what are the variables involved in such reform process? (iii) which variables have a positive or a negative impact on the implementation of decentralization processes? (iv) which variables could be affected by policy interventions, and how? This study aimed to answer these questions through the following objectives: (i) analyze the factors that have potentially affected the results of decentralization process in SSA basins, and (ii) analyze the performance of decentralization process in SSA basins.

As described in Chap. 1, these objectives were analyzed by combining qualitative analyses through a case study approach in three river basins (Limpopo in Mozambique, Inkomati in South Africa, and Mzingwane in Zimbabwe) in the SADC region, and quantitative analyses based on the data collected from twenty-seven river basin organizations in SSA countries.

Previous studies on the decentralization process of water management in Africa identified different factors that might have been limiting the decentralization of water management in SSA countries, such as the lack of clarity in terms of power relations and distribution of competences between central and local institutions and between old and new organizations, the insufficient financial sustainability of the managing agencies, the lack of knowledge and skills available to manage water at the various institutional and geographical scales, the conflicts arising from colliding interests, the unclear role of the state, the difficult public-private relations, the lack of reliable data and information, and cultural impediments.

Although past studies brought informative results regarding the limiting factors toward decentralization of water management in SSA countries, they are limited as they used qualitative approaches that did not estimate the directions and the magnitude of these factors on decentralization process and performance. To fill this gap,

J. Mutondo et al., *Water Governance Decentralization in Sub-Saharan Africa*, SpringerBriefs in Water Science and Technology, DOI 10.1007/978-3-319-29422-3_7

this study applied in SSA jointly qualitative and quantitative approaches, following the analytical and empirical framework developed and used by Kemper et al. (2007), and Dinar et al. (2007) to analyze water management decentralization. This framework described in Chap. 3, previously used in several regions of the world but not in Africa, was applied both to case studies (phase 1) and to the whole Sub-Saharan Africa (phase 2). Some modifications to the original framework were made to capture issues faced by water sector in SSA countries, such as the effect of climate change, as well as whether or not the basin in question is governed by an international river basin organization.

Chapter 4 applied the analytical framework described in Chap. 3 to summarize and compare the results about decentralization process and performance of water management in three river basins of SADC countries. Chapter 5 presents the empirical model used for the quantitative analysis in twenty-seven basins in SSA, and Chap. 6 illustrates the respective results.

The overall findings and conclusions from the study are presented, and their implications to water sector policy are discussed in this chapter. The conclusions and implications are given for the water management decentralization process and performance, taking into account the key variables of the analytical framework: (i) contextual factors and initial conditions, (ii) characteristics of the decentralization process, (iii) characteristics of central government/basin-level relationships and capacities, and (iv) the internal configuration of basin-level institutional arrangements.

7.1 Decentralization Process

7.1.1 Contextual Factors and Initial Conditions

Comparing the studied basins, high population density seems to yield pressure on basin resources that, in turn, foster the initiation of the decentralization process. This hypothesis was tested in the empirical analyses by inclusion of relative water scarcity variables in decentralization process models. The quantitative analyses showed that *Relative Water Scarcity, Share of Surface Water*, and *Water Flow Fluctuates* are significant and show a positive sign. This suggests that water scarcity in the range observed in our sample leads toward more involvement of the WUAs, more likelihood of creation of the RBO, and dismantling of existing institutions in the process of decentralization. The course of decentralization process is therefore more likely to be successful in settings with high populations, which leads to relative scarcity of water resources.

In terms of the level of economic development, our results showed that a higher political cost (a proxy variable for the level of economic development in the empirical analysis) increases the water users' involvement, and may lead to the creation of an RBO as a way to establish the new framework for a cooperative use of the resources. Additionally, under the performance models, basin budget per capita showed to be reducing basin problems after the decentralization process. The level of economic development contributes therefore positively in decentralization process of water management.

Finally, the results of the performance models indicated that the decentralization process is more likely to succeed in settings with less skewed distribution of basin resources, as basin stakeholders will be equipped with resources that allow them to cooperate and interact equally in the management of the basin resources.

7.1.2 Characteristics of the Decentralization Process

Descriptive statistics from the quantitative analyses revealed that decentralization processes in SSA countries, on average, are about one decade old, ranging between 2 and 30 years. Empirical analysis showed that as the number of years increase, the involvement of water-user associations in decentralization process decreases. This implies that, above a certain threshold, the number of years could contribute negatively to the decentralization process, as the stakeholders might be unwilling to continue the process if tangible results are not realized.

In terms of *type of devolution of the decentralization process*, results from the continent-wide study show a negative impact on the decentralization process in basins that followed a bottom-up approach. This is perhaps due to the fact that WUAs that have been established were not well prepared to implement the decentralization process, lacking organizational, legal, and technical skills. To confirm this fact, the level of managerial skills showed to be limited in the three studied SADC river basins. This implies the need of government support in terms of transfer of technologies and skills to manage water resources in SSA basins. The bottom-up devolution process is therefore a necessary but not sufficient condition for institutional decentralization needing support of government to transfer responsibilities and technical skills.

7.1.3 Central Government/Basin-Level Relationships and Capacities

The devolution of power to manage basin resources in the three studied SADC basins is still relatively low. The source of river basin budget is heavily skewed, being mainly from river basin stakeholders in Zimbabwe and from government and donors in South Africa and Mozambique.

Although in the three case studies the results showed a limited devolution of basin management activities to the basin level, empirical results in the continent-wide sample showed an increase in terms of participation of basin organizations in the management of basin management activities.

7.1.4   The Internal Configuration of Basin-Level Institutional Arrangements

The *presence of basin-level governance institutions* and a well-structured hierarchy of managing organizations can be seen as a potentially positive factor for water decentralization process and performance in the three SADC case studies. However, the power given to organizations located at basin level is limited. Additionally, mechanisms for information sharing and *forums for conflict resolutions* exist, but the participation of stakeholders is still limited. Our results from the African-wide survey showed finally that the likelihood of an RBO being created increases if the basin belongs to an international treaty.

7.2   Decentralization Performance

In this study, performance was measured by: (i) the RBOs level of success in attaining the objectives of decentralization of water management, (ii) the level of devolution of activities related to management of water resources, and (iii) the level of problems related to river basin stressed resources before and after decentralization process.

The results of empirical analyses showed that (i) the successes of decentralization process is more likely to be attained in the basins with institutional arrangements, following a bottom-up process, with uniform share of water resources and upon existence of financial resources to fund the process. Regarding (ii), the decentralization of water management in SSA countries has been implemented with some degree of transfer of basin activities from central government to basin organizations. The reduction of involvement of central government is significant for the activities related to water administration and enforcement of water quality. For (iii), decentralization of water management in SSA countries is contributing positively in reducing the constraints posed by different basin stressed resources. However, many problems due to water stress are still present and urgent to approach.

7.3   Policy Implications

Decentralization efforts in river basins have been seen around the world under various political and institutional situations. African river basins have been joining the decentralization process of river basins relatively late, initiating the process somewhere in early 1990s. After analysis, we conclude that the analytical framework of water management decentralization we used is robust enough to explain the decentralization process and progress even in the presence of a limited sample.

It seems that this framework, when used with a richer dataset and over a longer period of time, can be informative to policymakers when designing and evaluating decentralization processes in Africa and other parts of the world.

Some of the variables studied in our quantitative analysis have interesting implications. They reveal that the success and stability of the decentralization process depends on the way the new framework distributes the *Political Cost* and compensates those who carried its burden. As for the *Method of Creation*, a grass-roots initiative, despite all the benefits it may capture in terms of legitimacy and use of pre-existing community arrangements, is insufficient if not properly supported by government transfers of skills, or know-how, budget responsibilities, and technical knowledge. The similar impact of the variable *WUA's Involvement* in the presented model amplifies that conclusion. For SSA, this conclusion is probably the most relevant one, with policy implications. Training the WUAs prior to the initiation of the decentralization process is essential for a more effective decentralization process. Otherwise the social investment in institutional reforms in the water sector would be wasted. It should be mentioned here that the results concerning the variables *Method of Creation, Creation Bottom-Up*, and *WUAs Involvement* in a previous study with similar analytical framework applied to regions other than SSA were the opposite, suggesting that in SSA grass-roots efforts still have to be nourished.

Interpreting the opposite signs of the coefficients of major variables (*Creation Bottom-Up, Political Cost, Years Decentralization*) when they are included in estimates of the decentralization process on one hand and performance on the other hand could mean that while the implementation of decentralization processes in the water sector in SSA does not guarantee success, on the other hand, factors that improve the performance of decentralization do not necessarily facilitate its implementation. For example, in-progress decentralization institutions can have better results in terms of solving local water-related issues than established RBOs suffering from untrained staff and mal performance of infrastructure, and being disconnected from the stakeholders.

It also appears that the best performances of decentralized basins refer to solutions of infrastructural problems (floods, and land degradation control), while the socio-economic problems perceived before decentralization (conflicts, development) have been less addressed. This result could be a consequence of the fact that hardware solutions (infrastructure, engineering) are easier to implement than software solutions (stakeholders' participation, dispute resolution forums, etc.). Another interpretation of this last observation is associated with the previously mentioned context of un-trained staff: that infrastructure could be built by international companies, but when completed and left with local operators, may not function well due to inadequate institutions and preparedness.

References

Dinar, A., Kemper, K., Blomquist, W., & Kurukulasuriya, P. (2007). Whitewater: Process and performance of decentralization reform of river basin water resource management. *Journal of Policy Modeling, 29*(6), 851–867.

Kemper, K. E., Blomquist, W., & Dinar, A. (2007). *Integrated River Basin Management Through Decentralization*, World Bank and Springer.

Annex A
Application of the Analytical Framework to the Three Southern African Case Studies

The synthesis presented in Chap. 4 is based on case studies implemented in three water catchments of Mozambique, South Africa, and Zimbabwe. These catchments are the Mozambican portion of the Limpopo basins, the Inkomati, and the Mzingwane, respectively. Each case study described the characteristics of the catchment and the institutional variables, including their impacts on the decentralization process and performance (see Matsinhe et al. 2012; Chibwe et al. 2012; Musinakle et al. 2012). The following sections of this annex illustrate in detail the situation in the three studied catchments according to the variables that are identified in the analytical framework presented in Chap. 3.

A.1 Contextual Factors and Initial Conditions

A.1.1 Level of Economic Development of the Country and River Basin Before the Decentralization Initiative

In the Mzingwane basin, Musinake (2011) reports that the economic conditions of Zimbabwe are not favorable for the development of new institutional arrangements capable of implementing successful decentralized and integrated water resource management. The author underlines that the level of economic development in the catchment and in the country as a whole has been decreasing in the last decade. The treasury has been running dry given that the International Monetary Fund (IMF), the World Bank and other financial institutions had withdrawn their financial support to the government. In this respect, the government had started weaning off other responsibilities it felt were less strategic. At the same time, stakeholders were handicapped by hyperinflation. This situation made it impossible for stakeholders and government to invest time and money into knowledge generation, planning, negotiation, adoption, and implementation of institutions for river basin management, which have affected negatively the decentralization process.

In the Inkomati River basin of South Africa, in addition to the improved economic conditions over the past decade, the CMA has been receiving funds from the government and external donors. Especially, the funds received by the Inkomati catchment management agency (ICMA) have increased from about 5 million rands in 2006 to about 30 million rands in 2010. An increase in financial resources allowed the river basin agency to have financial capacity to bear transaction costs associated with decentralization initiative and ongoing costs that support and facilitate basin scale management.

The Mozambican portion of the Limpopo basin's GDP seems to be low, since the majority of basin stakeholders are smallholder farmers whose revenues from crop production are insufficient to cover the costs of water (Matsinhe 2011). Additionally, Mozambique had recently experienced a devastating civil war, which resulted in massive destruction of productive infrastructures and affected dramatically the economic development of the country. For example, between 1981 and 1986, the Mozambican GDP reduced by 30 % (Howard et al. 1998).

A.1.2   River Basin Population Density

The Mzingwane River basin has nearly 693,000 inhabitants in an area of 63,000 km^2, resulting in a population density of about eleven people per km^2. The same is observed in the Mozambican portion of the Limpopo (856,000 inhabitants within an area of 79,800 km^2). On the other hand, the Inkomati River basin has 2.2 million people in an area of 28,800 km^2, corresponding to a population density of about seventy-seven people per km^2. Dinar et al. (2007) report that the decentralization process is likely to be fostered in the basins with higher population density.

A.1.3   Stakeholders' Share of River Basin Resources Before the Decentralization Process

Musinake (2011) reports that the multiplicity of ethnicities and other deep socio-cultural differences among the Mzingwane catchment stakeholders throughout the basin has been a great challenge to establish communications and information sharing. Difficulties relating to differences in stakeholders' socio-economic status were increased by the type of devolution that followed the decentralization of the Mzingwane River basin. A top-down approach was followed in which the government of Zimbabwe solely decided to cede some powers to the stakeholders in water resources management by crafting two institutional arrangements, namely the Zimbabwe National Water Authority (ZINWA) and the Mzingwane Catchment Council.

In the Inkomati catchment, Chibwe (2011) reports that the distribution of river basin resources was highly skewed in favor of the minority of white South African citizens as heavy legacy of the apartheid regime, which only ended in 1994. South Africa has a Gini coefficient of 0.96, in terms of water use (Van Koppen et al. 2002). This statistic reveals a large gap between water use and the equity line, thus leaving many people without sufficient water resources for their daily usage. The inequality in accessing and using water resources is partly attributed to the poor state of some water infrastructure in the Inkomati Water Management Area (IWMA). Finally, formerly disadvantaged individuals, particularly in former homelands (Bantustans), continue to face significant power imbalances in terms of knowledge and expertise, compared to established white commercial farmers and other elite interest groups. There are differences between emerging farmers and commercial farmers in the IWMA in terms of water use. The commercial farmers, who are better endowed, are considered to be using more water than the quantity allocated to them, as they have been pumping water during non-pumping hours. In Mozambique, while not mentioning the socio-economic gaps of the two previous cases, Matsinhe (2011) reported a generalized low access to water resources by the local stakeholders.

A.1.4 River Basin Stakeholders' Management Capacity

In the Mzingwane catchment, capacity building programs were not reported, but sufficient human capacity seems to exist. This capacity is demonstrated according to Musinake (2011) by the ability of the catchment and sub-catchment councils to prepare the outline plan for the basin.

In the Inkomati catchment area of South Africa, Chibwe (2011) reports that the CMA has built its managerial capacity over the period of its existence and it is now able to offer services to other CMAs. For example, the Inkomati CMA has produced the catchment management strategy and has been invited by the Breede Overberge (BO) CMA to provide input into its drafting of the basin organization catchment management strategy. In Mozambique, according to Matsinhe (2011), the Limpopo basin is just an example of the generalized lack of human capacity and resources for water management observable all over the country. Similar to South Africa, capacity building was not reported in Mzingwane River basin, although the majority of basin population did not complete primary school. However, human capacity seems to exist as the basin stakeholders were able to prepare the outline of the river basin plan.

A.2 Characteristics of Decentralization Process

A.2.1 Length of Decentralization Process

In the Mzingwane River basin, the process has been underway for eleven years, since the creation of ZINWA in 2000. The length of time needed to complete a decentralization process is difficult to assess, and there is a need for adequate time to adjust changes and stabilize the decentralization process (Blomquist et al. 2005). Therefore, the direction of the decentralization process cannot be easily assessed using the number of years that Mzingwane River basin has been under decentralization.

In the Inkomati WMA, according to DWAF (2001), the establishment of the Inkomati CMA was initiated in July 1997 by the regional office (RO) of DWAF Mpumalanga. On the 30th of March 2004, the Inkomati CMA was officially launched. It took almost seven years since the approval of water law in 1998 to establish the ICMA. In the Mozambican portion of the Limpopo River basin, the decentralization of water resource management started with the approval of the Water Law in 1991, which resulted in the establishment of the river basin organization (Limpopo River basin management unit) in 1993. The decentralization process has been therefore underway for almost eighteen years, and it is still an ongoing process.

A.2.2 Number of Institutions Created or Dismantled During the Decentralization Process

Musinake (2011) reports that decentralization of water management in Zimbabwe eliminated and created institutions at central and local levels. Specifically, at the national level, the Ministry of Water Resources Management and Rural Development, as well as ZINWA, were created while the Department of Water and Development was dismantled. At the local level, district offices and structures of Department of Water and Development were dismantled, while the Mzingwane catchment and sub-catchments, such as Sashe, upper Mzingwane, Lower Mzingwane, and several water-user associations were created. Each catchment and sub-catchment is led by a council.

In South Africa, the decentralization process did not eliminate existing institutions at the national level, while it created and eliminated local-level institutions. The Inkomati catchment management agency was established, and two irrigation boards were converted into water-user associations.

In the Mozambican portion of the Limpopo, at the national level, the national directorate for water and regional water management agencies were created. At the river basin level, the decentralization process has created the Limpopo River basin management unit (UGBL),[1] the Chokwe hydraulic public enterprise (HICEP), the Baixo Limpopo Irrigation Scheme (BLIS),[2] the Basin Committee[3] and some

---

[1]UGBL is a river basin organization under the management of ARA-Sul, which is responsible for water allocation at the basin level.
[2]HICEP and BLIS are public enterprises responsible for the management of irrigation schemes in Chokwe and Xai-Xai districts, respectively.
[3]The Basin Committee is a coordinating organ between the entity responsible for water allocation and other river basin stakeholders.

water-user associations.[4] However, Matsinhe (2011) reports that the existing water user associations are not fully operational. The limited functionality of the water user associations is also reflected in the lack of formal inclusion of this type of organization in the management structure of the river basin organization.

A.2.3 Level of Involvement of the River Basin Stakeholders in the Decentralization Process

In the Mzingwane River basin, the only stakeholders who actively participate in crafting water laws and creating river basin organizations are government officials and politicians. Specifically, Musinake (2011) reports that the government unilaterally made the decision to form the ZINWA, and the local stakeholders were never consulted in the promulgation of the ZINWA Act of 1998.

In South Africa, different stakeholders were involved in the development of the 1998 Water Act, as well as in the creation of river basin organizations. At IWMA, the involvement of stakeholders was led by the government through the DWAF regional office (RO) in Mpumalanga and the process started in 1997 before the approval of the 1998 Water Act. The identified stakeholders were either contacted by phone or mail by DWAF officials. Each time new stakeholders were identified, they were also contacted and motivated to participate in the proposal development process for the establishment of the Inkomati CMA. In order to guarantee the participation of disadvantaged stakeholders, DWAF officials traveled to historically disadvantaged communities and companies to hold meetings with them. In cases where participants had incurred transportation costs, they were reimbursed by the government through the DWAF RO (DWAF 2001). When the 1998 NWA was passed each sub-catchment of the IWMA (Komati, Crocodile, Sabie-Sand) developed a sub-catchment proposal. Finally, the three proposals were amalgamated in 2000 to form a CMA (Inkomati CMA) proposal that was submitted to DWA for consideration and approved in 2001. These results show strong participation by stakeholders in the creation of ICMA and its sub-catchments.

In the Mozambican portion of the Limpopo, the creation of river basin organizations (ARA-Sul and UGBL) was mainly performed by government officials in response to World Bank and other funding agencies recommendations. Matsinhe (2011) reports that formal basin management institutions, such as UGBL, HICEP, and BLIS were created by the government and, in part, through national laws and decrees. In addition, communities have a smaller share of responsibility in the basin management issues.

Participation of stakeholders on decentralization process can also be measured by *the composition of sub-catchment councils*. Musinake (2011) finds that in all

---

[4]Matsinhe (2011) reports that sixty water-user associations have been created in Limpopo River basin.

sub-catchment councils, female representation is less than 40 %. Additionally, there is no single woman who heads any of the sub-catchment councils. The highest position for a woman is the treasurer, which is registered at Upper Mzingwane sub-catchment council.

The participation of stakeholders in the decentralization process can also be demonstrated by the level of involvement of local stakeholders in ZINWA committees. Most of the interviewed individuals stated that local stakeholders are not involved in ZINWA committees.

When the Inkomati CMA was formally established in 2004, its capacity was low with a lean staff structure and no governing board in place. The board was appointed in 2006 to oversee the operations of the Inkomati CMA. The governing board of ICMA was initially composed of thirteen members, representing different stakeholders[5]; however, during the period of data collection, the board size was ten members only.

In the Mozambican portion of the Limpopo, UGBL, HICEP, and BLIS governing board members are appointed by the government. And water-user associations governing board members are appointed by the local stakeholders, using a voting system. The government power at river basin level is also highlighted by the governing body of Basin Committee. Matsinhe (2011) reports that the basin committee is chaired by the director of the UGBL, a representative of ARA-Sul, which is an organization related to the central government.

The level of stakeholders' participation in the decentralization process can also be measured by *the degree of participation of stakeholders in river basin meetings*. In Mzingwane river basin, 75 % of stakeholders have been participating in river basin meetings. However, the usage of English language and western protocol has limited the participation of stakeholders during basin meetings.

In the Inkomati catchment, the level of attendance to board and basin meetings were reported to be 100 and 80 %, respectively. Although the majority of basin stakeholders attend the basin meetings, it was made clear by the respondents that some of the members of the governing board of the Inkomati CMA were passive and did not participate fully in the board deliberations. Most of the members who were alleged to be silent during most board meetings are those that represented disadvantaged communities of former homelands.

In the Mozambican part of the Limpopo River basin, information-sharing and communication among basin stakeholders occur mostly through meetings. Although basin meetings are the main mechanisms used for sharing information, the survey respondents were not able to estimate the level (in percentage) of stakeholders' participation. They reported that there is a good stakeholders' attendance of river basin meetings but the decisions are mainly taken by the basin committee, which is presided by ARA-Sul. Matsinhe (2011) reports that small

---

[5]Chibwe (2011) reports that each of the following stakeholders (industry, mining, and power generation; commercial agriculture; civil society; tourism and recreation; productive use of water by the poor; forestry; conservationist; traditional leaders; and SALGA) have a representative in the boards. The remaining members represent government agencies.

farmers and water-user associations located remotely from the decision centers in the Limpopo basin are virtually excluded, and have non-meaningful participation in the decision-making process. The same authors indicate that farmer associations that are located far from the urban centers where meetings take place have claimed that they are not invited to participate in the basin committee meetings, and for others it is difficult to participate in the meetings due to the associated costs of accommodation and transport.

A.2.4   The Type of Devolution Used in the Process of Decentralization

Finally, interpreting the results of the three case studies, the decentralization of Mzingwane River basin and in the Mozambican portion of the Limpopo basin followed top-down devolution, while in the Inkomati the process initially started as a top-down approach led by the DWAF Regional Office in Mpumalanga; however, it turned out to be a mutually desired process, when stakeholders joined the process.

A.3   Central Government/Basin-Level Relationships and Capacities

A.3.1   Percentage of Tariffs Remaining at the Basin

In the Mzingwane River basin, 75 % of tariffs stay in the basin and the remaining 25 % is channeled to the central government as value-added tax. Musinake (2011) reports that of all the revenues generated within the basin, stakeholder organizations get much less than 1 %, while ZINWA, the statutory authority, collects 74 % of revenues from water tariffs.

In the Inkomati Water Management Area, according to Chibwe (2011), none of the water tariffs are managed by stakeholders and, therefore, the Inkomati CMA does not have financial autonomy, and it is heavily dependent upon external donors and the government financial resources to finance basin activities.

In the Mozambican portion of the Limpopo, none of water tariffs collected remain at the basin level. Regarding the possible destinations of water revenues, according to regulation of water tariffs (Decree 43/2007), the government retains 100 % of the collected revenues from water tariffs with the following distribution: 40 % of tariff revenues go to Ministry of Finance, and 60 % to Ministry of Public Work and Housing.

The main *source of river basin budget* in the Mzingwane River basin were river basin stakeholders, representing 98 % of the river basin budget, while 1 % is from the government, and the remaining 1 % is from nongovernmental organizations

(NGOs). The fact that the majority of river basin resources are from river basin stakeholders might guarantee financial sustainability over time. However, the low contribution from the government might indicate the lack of government commitment in the decentralization initiative. In the Inkomati, no reliable data was collected on the source of budget. While budgetary autonomy is one of the main principles of CMA relations in South Africa, Chibwe (2011) reports that the Inkomati CMA has currently two funding profiles: A parliamentary allocation that comes from government coffers, and an external funding that comes from the donor community.

A.3.2 Level of Authority Held by River Basin Stakeholders on Managing River Basin Resources

In the Mzingwane catchment, a shift of the function of water resources allocation from the water courts and direct government control into the hands of ZINWA, and Mzingwane catchment council, a stakeholder institution, was a step toward decentralization. Additionally, the abolishment of the water rights system in favor of renewable water permits has been a catchment-based form of water allocation. In this regard, all the water permits issued within the basin have been issued by the stakeholder organizations. In the Mozambican portion of Limpopo, the majority (61.9 %) of respondents reported that the river basin organizations do not have the necessary authority/independence in managing water resources.

A.3.3 The Level of Authority Given to Different Stakeholder Groups to Manage River Basin Resources Before and After the Decentralization Process

In the Mzingwane catchment, results show that the responsibilities regarding infrastructure financing, setting water standards and water quality enforcement are still concentrated within the central government, because 100 % of the responsibility has been given to the national/central government since initiation of the decentralization process. Responsibility regarding water administration was shared by local and provincial government levels before the decentralization process. The decentralization process improved the participation of river basin stakeholders in management of water administration. Specifically, 75 % of water administration responsibility was given to river basin stakeholders, and the remaining 25 % was given to local level government.

Although local-based organizations have been involved in water management, Musinake (2011) indicates that the involvement of stakeholders in decision-making is marginal, as ZINWA is the supreme body that makes all decisions relating to water in the catchment. The Mzingwane catchment council and its constituent at the four sub-catchment councils are mainly restricted to housekeeping issues. They are only involved in preparation of the catchment outline plan, monitoring water flows and data collection, and in some part, a conduit for water levies from water users to the national authority. Evidently, the distribution of power and authority and discretion over the use of water-related revenues is highly skewed toward the national authority.

In the Mozambican portion of the Limpopo, all survey respondents reported that responsibilities regarding infrastructure financing, water-quality enforcement and setting water standards are performed by the central government. Additionally, Matsinhe (2011) points out that the weak authority given to local organizations can be highlighted by the level of authority shared by the government and local-level organizations in the management of water infrastructures. Operational management of the hydrological resources at the Mozambican portion of Limpopo River basin is performed by ARA-Sul, an organization that is subordinated to the national directorate of water (DNA), a government-controlled unit. Existing infrastructure, like the Chokwé and Xai-Xai Irrigating Schemes, were transferred from central government control to the Chokwe hydraulic public enterprise (HICEP) and Baixo Limpopo Irrigation Scheme (BLIS), respectively. As both HICEP and BLIS are subordinated to the government through the Ministry of Agriculture, this fact suggests that river basin management tasks are mainly performed by related governmental institutions with weak participation of local representatives.

A.4 Configuration of Basin-Level Institutional Arrangements

A.4.1 River Basin Organizational Structure, Including the Composition of Each Organ and its Function

In the Mzingwane catchment, there are two existing main structural arrangements: the catchment councils, and ZINWA. The water authority's affairs are run by a ZINWA board, which is composed of ten members. It is worth noting that issues of policy and high-level decision-making relating to water resource management are deliberated at ZINWA's level. What matters most is how this board has been constituted. According to Musinake (2011), the state through the minister of water resources appointed the board chairman, the chief executive officer, and the four other board members. In addition to the board members representing state interests, the responsible minister chose the final four members of the board from a list of five prospective members, forwarded by the catchment councils (Musinake 2011). This autocracy has found its way to the lowest level, as well. Musinake (2011) points out that ZINWA officials, especially at the onset of the decentralization initiative, appointed themselves as the sole stakeholders privileged to elect representatives in the sub-catchment councils.

However, Musinake (2011) notes that the current establishment of water-user organizations in the Shashe sub-catchment represents a step in the right direction.

These organizations certainly will enjoy some sort of recognition and support from the non-government organizations (NGOs), provincial and district-level state structures, as well as from research institutes.

Chibwe (2011) reports that the Inkomati Water Management Area is governed by the Inkomati Catchment Management Agency (ICMA). The ICMA is lead by a governing board; however, the daily activities of the ICMA are lead by a chief executive officer (CEO) assisted by managers and support staff. The ICMA interacts directly with three executive committee officers representing the three sub-catchments (Sabie, Crocodile, and Komati). Below the executive committees are representatives of water users. River basin water users are organized in associations (water-user associations) and irrigation boards. The irrigation boards are in the process of being transformed into water-user associations (WUA). However, only two irrigation boards have been formed. The functionality of the WUAs is still weak, since only one WUA is currently functional.

The Mozambican portion of the Limpopo River basin is under the management of ARA-Sul. The implementation of water related strategies and policies is particularly led by the Ministry of Public Works and Housing, which is organized in directorates. The directorate responsible for water resource management is the National Directorate for Water (DNA), which coordinates the activities of the five regional administrative offices (ARAs). Under the decentralization process, operational management of the hydrological resources at the Mozambican portion of the Limpopo River basin was given to ARA-Sul. ARA-Sul responds to DNA but it has financial and administrative autonomy. At the Mozambican portion of the Limpopo River basin, ARA-Sul is represented by the Limpopo River basin management unit (UGBL). UGBL works like a section within ARA-Sul, and it is responsible for implementing the general scheme of water use at the basin level, and ensuring that existing water resources meet existing demand.

The involvement of river basin's stakeholders in the management of water resources at the basin level is done through the Limpopo River basin committee. The basin committee is chaired by the UGBL director, a staff member of the ARA-Sul, and it is composed of different stakeholders, including representatives of the private sector, water-user associations, the Chokwé and Xai-Xai irrigation system managers, religious institutions, farmers, and representatives from other economic and political sectors. Despite the presence of local stakeholders in the decision-making entities, Matsinhe (2011) reports that that UGBL and the Limpopo River basin committee implement central government policy at the basin level, and the community members have a small share in water management authority and responsibilities. Other water users are represented by water-user associations. The Mozambican portion of the Limpopo River basin counts actually about sixty water-user associations, of which thirty-two have been legally registered. The internal configuration of the Mozambican portion of the Limpopo River basin shows evidently the effort of decentralizing the management of river basin resources. However, the predominance of government-created institutions with weak involvement and functionality of basin-based organizations is still evident.

A.4.2 Information Sharing

In the Mzingwane River basin, information-sharing takes place basically through basin meetings. However, calendars, annual reports and strategy documents are becoming increasingly effective mechanisms as people change their attitudes. Musinake (2011) reports that information sharing through basin meetings has been carefully crafted to decrease the participation of local stakeholders. For example, in council meetings in the Mzingwane catchment, foreign language (English), Western protocols, and practices have been observed and held with high esteem against a background of a less-literate audience. Alien language has been ensuring that the interests of scientific, political, and commercial stakeholders are prioritized, while the majority of traditional leadership and communal interests are sacrificed.

In the Inkomati, WMA respondents reported that there are different mechanisms for information sharing, such as board meetings, annual reports, and radio broadcasts. Basin reports and profiles are also used as mechanisms for information sharing. However, the limited active participation of basin stakeholders (mainly disadvantaged groups) in basin meetings contributes negatively to decentralization process.

All (100 %) case study survey respondents reported that the Mozambican portion of the Limpopo River basin has forums for information sharing. Communication among members of the same association takes place mostly through meetings. Additionally, survey respondents reported that within water-user associations, meetings are scheduled on weekly bases, and the meetings among associations are scheduled on an irregular basis, depending on the occurrence of problems in the basin. HICEP and the BLIS are invited to participate in meetings organized by water-user associations. Interviewees indicated that UGBL and water users away from the Chokwé and Xai-Xai irrigation systems meet infrequently. At the basin level, the basin committee meets twice a year, while HICEP, BLIS, and the water-user associations meet on monthly bases.

A.4.3 Forums for Conflict Resolution

The Mzingwane River basin has seven forums for conflict resolution, namely Mzingwane catchment council and its four sub-catchment councils, ZINWA, and the Ministry of Water Resources Management and Rural Development. However these forums do not effectively solve river basin conflicts. According to Musinake (2011), developments in the Shashe sub-catchment have revealed that basin stakeholders have been denied a forum to get their voices heard by water authorities. In particular, the army and police have used force and intimidation to get their interests across.

In the Inkomati water management area, respondents reported that legal arrangements (water tribunals) exist, which have been effectively used for water conflict resolutions. In the Mozambican portion of the Limpopo, 100 % of the survey respondents pointed out that the Mozambican portion of Limpopo River basin does not have forums for conflict resolution at the basin level. The basin committee works slightly like a forum to hear disputes, when called for, but without authority to solve them.

A.5 Performance Assessment

The following results show how interviewees interpret the performance of the newly established institutions with respect to a number of water management problems before and after decentralization. Respondents were asked whether, to their knowledge, selected issues existing before decentralization have improved or worsened after the process.

A.5.1 Level of Accomplishment of the River Basin Objectives

The main objectives of Mzingwane River basin decentralization process reported by ZINWA officials were reduction of water conflicts and the improvement of equitable allocation of water permits. The majority (66 %) of respondents of the semi-structured questionnaire also reported that the main objective of the Mzingwane River basin decentralization process was to improve water allocations. ZINWA officials reported that while decentralization has decreased water conflict problems by 75 %, it did not improve water allocation. The limited success in improvement of water allocation is also supported by respondents of semi-structured questionnaire, as the majority (60 %) reported weak improvement of water allocation in the catchment.

In the Inkomati WMA, the main objectives of the decentralization process were to reduce water scarcity and water conflicts, as well as assuring water quality. Survey respondents indicated that these objectives have been reached partially. The decentralization process improved by 25 % the problems related to water scarcity and conflicts, and by 50 % the problems related to water quality. These results suggest that there are signs of improvement in performance of the decentralization process in the Inkomati water management area.

In the Mozambican portion of the Limpopo, survey respondents reported that the main objectives of the UGBL are to improve water allocation and distribution (85.7 %), and crop production (14.3 %). The respondents were not able to assess the level of accomplishment of these objectives. However, Matsinhe (2011) reports that water allocation is still poor, due to lack of improved water distribution infrastructures, and crop production is also still low. These findings suggest that the main objectives of the UGBL are still far to be attained.

A.5.2 Level of Problems Related to River Basin Stressed Resources Before and After Decentralization Process

Respondents were asked to rank the level of problems associated with the river basin's stressed resources before and after the decentralization process using the following categories: (1) no response, (2) no problem, (3) some problem, and (4) severe problems. The evaluated stressed resource problems were: water scarcity, floods, environmental quality, land degradation (erosion, salinity, etc.), water conflicts, water storage, and river ecology, among others.

The ZINWA officials indicated that the decentralization process of Mzingwane River basin reduced the problems related to water scarcity, water conflicts, and water conservation and storage. While water scarcity and conflicts were considered problems before decentralization, they were not considered to be a problem after decentralization. However, decentralization increased problems related to river ecology and land degradation. Similar to the results reported by ZINWA officials, the results from the 117 semi-structured questionnaires submitted to river basin stakeholders reveal that decentralization decreased problems of water scarcity, and increased problems of environmental quality and soil erosion.

In the Inkomati WMA, the decentralization process did not change the state of the majority of the problems listed above, as they were mostly ranked to have some problems (category 3) before and after the decentralization process. However, the decentralization process improved the availability of water and reduced water conflicts. Both problems shifted from category 4 (severe problem) to category 3 (some problems). These results indicate that decentralization performance has been increasing according to the respondents.

In the Mozambican portion of the Limpopo, the majority of respondents consider that the conditions of stressed resources before the decentralization process are severe, and they do not improve substantially after decentralization, which indicates the low performance of the process in the Mozambican portion of Limpopo River basin.

Respondents were finally asked to report the existence of water rights before and after the decentralization process, and which takes responsibility for awarding water rights, water allocation, modeling and forecasting water availability, monitoring and enforcing water quality, and collecting water tariffs before and after the river basin decentralization process.

ZINWA officials indicated that permanent water rights prevailed before the decentralization process, and they were replaced by water permits renewable after two years through the decentralization of Mzingwane River basin. They also reported that responsibility regarding water allocation, modeling, and forecasting water availability and collecting tariffs was given to state/provincial government agencies before the decentralization process, and it is now performed by the river basin authority (ZINWA).

In the Inkomati WMA, respondents indicated that before the introduction of the new NWA and subsequently the decentralization initiative, there were permanent water rights, and these rights were eliminated with the introduction of the new NWA.

Finally, in the Mozambican portion of the Limpopo, all respondents reported that water resources belong to the state, and the rights to use are given by the state. However, after the decentralization process, the basin committee has also been responsible for water allocation and assigning water use rights. Water quality standards are set by the ministry of environmental coordination (MICOA), and water quality along the Limpopo River is monitored by ARA-Sul, along with MICOA. Monitoring the Limpopo River's flows in order to anticipate and identify flooding or insufficiency of water is under the responsibility of DNA, ARA-Sul, and the national institute for disaster management (INGC) through the emergency operative center (CENOE).

References

Blomquist, W., Dinar, A., & Kemper, K. (2005). *Comparison of institutional arrangements for river basin management in eight basins.* World Bank policy research working paper #3636, Washington, DC.

Chibwe, T., Bourblanc, M., Kirsten, J., Mutondo, J., Farolfi, S., & Dinar, A. (2012). *Reform process and performance analysis in water governance and management: A case of study of Inkomati Water Management Area in South Africa.* Working paper.

Department of Water Affairs and Forestry (DWAF). (2001). *Proposal for the establishment of a catchment management agency for the Inkomati basin.* For submission to the Minister of Water Affairs and Forestry. Mpumalanga, South Africa

Howard, J. A., Jeje, J. J., Tschirley, D., Strasberg, P., Crawford, E. W., & Weber, M. T. (1998). *What makes agricultural intensification profitable for Mozambican Smallhorder?* MSU International Department of Agricultural Economics, paper # 69, Michigan, USA, 124 pp.

Matsinhe, M., Mungatana, E., Mutondo, J., Farolfi, S., Dinar, A., & Tostão, E. (2012). *Reform process and performance analysis in water governance and management: A case of study of Limpopo river basin in Mozambique.* Working paper.

Matsinhe, M. P. (2011). *An assessment of the factors affecting decentralization performance in the management of water resources in Limpopo Basin (Mozambique).* MSc Thesis. University of Pretoria.

Musinake, G. (2011). *Reform process and performance analysis in water governance and decentralization: A case of Mzingwane Catchment in Zimbabwe.* MSc. Thesis. CASS, University of Zimbabwe, Harare.

Musinakle, G., Dzingirai, V., Mutondo, J., Farolfi, S., & Dinar, A. (2012). *Reform process and performance analysis in water governance and management: A case study of Mzingwane catchment in Zimbabwe.* Working paper.

Van Koppen, B., Jha, N., & Merrey, D. J. (2002). *Redressing racial inequities through water law in South Africa: Revisiting old contradictions?* Comprehensive assessment, Research paper 3. [Online] Available from http://publications.iwmi.org/pdf/H030391.pdf. Accessed May 12 2010.

Annex B
Major River Basins and River Basin Organizations in Sub-Saharan Africa

The quantitative analysis of this study is based on a sample composed of twenty-seven RBOs, located in six countries, distributed over two of the four SSA regions (four RBOs in two Eastern African region countries, and twenty-three RBOs in four Southern African region countries). The reasons for the use of this sample are described in Chap. 5. As the surveyed catchments represent only partially the situation in SSA in terms of water governance decentralized institutions, it is useful to present an overview on the major basins and basin organizations in SSA countries. This Annex responds to this need and strives to contextualize our quantitative analysis within the African landscape.

B.1 Major Water Basins in Sub-Saharan Africa

The African continent is composed of over fifty river basins, spanning nearly all its countries, some of which are international and some are domestic in nature. Among these, the major basins are Senegal, Volta, Niger in West Africa, Lake Chad, Ogooue, and Congo in Central Africa, Nile, Lake Turkana, Juba Shibeli in East Africa, and Zambezi, Okavango, Limpopo, and Orange in Southern Africa (Fig. F.2). The United Nations Economic Commission for Africa—UNECA (2000) adds from the list presented above the following river basins: Gambia, Sassandra, Comoe, Gueme, and Sanga in West Africa, Ogur in Central Africa, Awash, Omo, Tana, Pangani, and Rufuji in East Africa, as well as Kunene, Rovuma, and Save in Southern Africa.

This annex describes river basins affecting water flows in the Sub-Saharan region of the African continent. The differences in terms of socio-economic conditions, which determine the level of decentralization process and management of river basin resources, justify the separation of Northern Africa from the Sub-Saharan region of Africa. Hence, the following sections of this annex describe the main features of the major basins affecting water flows in different regions of Sub-Saharan Africa.

© The International Bank for Reconstruction and Development/The World Bank 2016   69
J. Mutondo et al., *Water Governance Decentralization in Sub-Saharan Africa*,
SpringerBriefs in Water Science and Technology, DOI 10.1007/978-3-319-29422-3

B.2   Major Basins in the Western African Region

Senegal basin: The Senegal River basin is estimated to cover an area of 483,180 km^2 and spread over four countries (Guinea, Mali, Mauritania, and Senegal). The basin rainfall varies from 55 mm/year in the valley and delta to 2000 mm/year in the upper basin in Guinea, with an overall basin average of 550 mm/year. The irrigation potential of Senegal basin is estimated to be as high as 240,000 ha in the Senegal River valley. In Mauritania, the irrigation potential of the Senegal basin is estimated to be as high as 125,000 ha. The total irrigation potential of the Senegal basin is estimated to be 420,000 ha. However, only 118,000 ha is presently under irrigated agriculture.

    Volta basin: The Volta River basin is shared by Burkina Faso, Benin, Cote d'Ivoire, Ghana, Mali, and Togo. The majority (85 %) of the river basin area is shared by Burkina Faso and Ghana. The basin covers 67 and 64 % of Burkina Faso and Ghana land mass, respectively. Rainfall in the basin ranges from 400 mm/year in the North to 1800 mm/year in the coastal zone and with evaporation of about 2500 mm/year. The irrigation potential of the Volta basin is estimated to be 142,000 ha.

    Niger basin: It is the second longest river in Africa after the Nile with about 4100 km long, and basin area is estimated to be 1,471,000 km^2. The basin spreads over in the following countries: Algeria, Benin, Burkina Faso, Cameroon, Chad, Cote d'Ivoire, Guinea, Mali, Niger, and Nigeria. Niger basin area covers about 7.25 % of the African continental landmass. The Niger basin is composed of the Niger River, which originates from Guinea with its tributaries of Bani, Gouroval, Dargol, Sirba, Gouroubi, Diamamgou, and Tapoa, all originating from Burkina Faso. The other tributaries include Mekrou, Alibori, and Sota, originating from Benin and Benue from Chad. The rainfall in the basin varies from 1200 to 3000 mm/year in Guinea zone, 500–1200 mm/year in Sudanese zone, and 100–500 mm/year in Sahelian zone. The total irrigation potential of Niger basin is estimated to be about 2,816,510 ha, while the present irrigation area has been estimated at 924,620 ha. Specifically, Niger River basin irrigation potential is estimated to be 1,678,510 ha in Nigeria, 556,000 ha in Mali, and 300,000 ha in Benin. The Niger basin has high hydropower potential of about 30,000 GWH with a current installation of 7000 GWH.

B.3   Major Basins in Central African Region

Lake Chad basin: The Lake Chad basin is located in Northern Central Africa and it covers almost 8 % of the continent and spreads over seven countries (Chad, Niger, Nigeria, Cameroon, Algeria, Sudan, and Central Africa). The total basin area is about 2,381,635 km^2 and the conventional area (20 % of total area) is about 427,500 km^2 from which 42 % is in Chad, 28 % in Niger, 21 % in Nigeria, and 9 %

in Cameroon. The basin has irrigation potential of about 2.0 million ha from which only 113,296 ha are actually under irrigation.

Ogooue basin: The Ogooue catchment area is estimated to be 223,856 Km^2 of which 173,000 Km^2 (73 %) lies within Gabon, and the remaining area is shared by Cameron and Congo Brazzaville. The basin is located in the equatorial region and the average rainfall is no less than 2.0 m with 2–3 dry months per year. Annual evapotranspiration in the basin is estimated to range from 1000 to 1250 mm per year.

Congo basin: It is the largest basin in Africa, and the second largest in the world next to the Amazon river basin. The Congo catchment area is estimated at 3.7 million km^2 and is shared by nine countries, namely: the Democratic Republic of Congo, the Central African Republic, the Congo (Brazaville), Angola, Cameroon, Burundi, Rwanda, Tanzania, Zambia, and Malawi, with the largest basin area being in the Democratic Republic of Congo. The Congo River basin consists of Congo River itself, its tributaries of Oubangui, Kasai, Sangha, Kuilu, Kwango, Ruki, Lamami, Lulonga, Amwini, and other smaller rivers. The average rainfall in the basin ranges from 1200 mm to more than 2000 mm in the center. Congo Catchment has a potential for irrigating 9,800,000 ha and it is actually irrigating 35,767 ha. The Congo basin has a hydropower potential of 39,000 MW at Inga with only 1775 MW installed. Additionally it still has large potentials for navigation, fishery and Eco-Tourism.

B.4   Major Basins in the Eastern African Region

Nile basin: It is the longest in Africa and second longest in the world. It flows through 6700 km from Egypt to Tanzania. The Nile catchment is estimated to be over 3 million km^2 (approximately 10 % of total land surface of African continent) and it covers the following countries: Burundi, the Democratic Republic of Congo, Rwanda, Tanzania, Kenya, Uganda, Ethiopia, Eritrea, Sudan, and Egypt. The mean annual rainfall over the entire basin is about 2000 billion m^3. The irrigation potential of the Nile basin is immense. For example, it has a potential to irrigate 4.8 million ha in Egypt, 200,000 ha in Uganda, and 300,000 ha in Eritrea.

Lake Turkana basin: The total basin area is about 130,860 km^2. This basin is mainly fed by the Lake Turkana, which is situated in the Great Rift Valley in the northwestern part of Kenya. Lake Turkana is situated in an arid and hot area with mean annual rainfall less than 250 mm. The main tributary of the basin is the River Omo, which contributes more than 90 % of the total water influx. The second largest river is the Turkwel River and the other rivers are temporary, flooding only during sporadic rains. The evaporation rate has been estimated at 2335 mm per year at the basin. The main activity in the basin is agriculture (pasture with about 47 % of basin area).

Shebelli—Juba basin: The catchment area is shared by Kenya, Ethiopia, and Somalia, and covers a total area of over 810,000 km^2 and more than 46 % of the basin is within Ethiopia. The total rainfall varies from 200 to 1800 mm/year with an average of 430 mm/year. The potential irrigation in the basin is estimated at 323,000 ha; however, less than 200,000 ha is currently under irrigation.

B.5 Major Basins in the Southern African Region

Zambezi basin: It is the fourth largest basin (catchment) area in Africa with a basin area of over 1.3 million km^2 and covers eight countries, namely: Angola, Botswana, Zambia, Zimbabwe, Tanzania, Malawi, Mozambique, and Namibia. Similar to the Nile basin, the irrigation potential of the Zambezi basin is immense. The annual rainfall in the basin varies from 1800 mm/year in the north to 550 mm/year or less in the south of the basin. The total irrigation potential of the Zambezi basin, spreading over eight riparian countries, is estimated at 3,160,380 ha, of which less than half is presently under irrigated agriculture. Specifically, it is estimated to have a potential to irrigate 1.7 million ha in Mozambique, 700,000 ha in Angola, and 422,000 ha in Zambia. The Zambezi basin has significant hydropower potential with an installed capacity of 4620 MW, and about 40 more sites with a total capacity of 13,000 MW identified.

Okavango basin: It covers about 725,000 Km^2 (approximately 1 % of African land mass), and it is shared by three countries, namely: Angola, Namibia, and Botswana. The rainfall in the basin ranges from 1300 mm/year in Angola to 300 and 400 mm/year in Namibia and Botswana, respectively. The irrigation potential of the Okavango basin has been estimated at 200,000 ha in Namibia and 600 ha in Botswana.

Orange basin: It is shared by Botswana, Namibia, Lesotho, and South Africa and covers almost 3 % of the continental landmass with an estimated area of about 896,368 km^2. The rainfall in the basin varies from 35 to 1000 mm/year with a mean value of 325 mm/year over the basin. The irrigation potential of the Orange River basin is 25,000 ha in Namibia, 12,500 ha in Lesotho, 352,500 ha in South Africa.

Limpopo basin: The Limpopo basin is shared by Botswana, Zimbabwe, South Africa, and Mozambique and covers an area of almost 402,000 km^2, over 46 % of which is in South Africa. This basin is composed of the Limpopo River and its tributaries, the Shashi and Elephant rivers. The rainfall in the basin ranges from 290 to 1040 mm/year with a mean of 530 mm/year. The irrigation potential of the Limpopo basin in South Africa is estimated at 131,500 and 148,000 ha in Mozambique. The overall total irrigation potential for the Limpopo basin across the four riparian countries is estimated at 295,500 ha, while the area under irrigation at present is about 242,000 ha.

As described above, the major river basins in Africa are international river basins, as they cover more than one country. Some of these river basins have set an international framework aiming to governing the management of river basin resources. The next section describes the major river basin organization in Sub-Saharan Africa.

B.6 International River Basin Organizations (RBOs) in Sub-Saharan Africa

The management of river basins described above had been mainly centralized and controlled by the government. However, in the past five decades, the world experienced changes in the management of water resources. These changes were mainly based on attempts to replace the centralized management approach with the integrated water resource management (IWRM) approach. IWRM gained acceptance after the International Conference on Water and Environment in Dublin in 1992. One of the main principles of IWRM is the decentralization of river basins management through the creation of river basin organizations.

Additionally, disputes among countries sharing the same basins and the need to implement development projects at the basin level following IWRM principles motivated the creation of river basin organizations in form of commissions, committees, and other organizational set-ups. In Africa, these organizations have been created at international and national levels with the following goals: (i) Development of management and action plans, (ii) monitoring water flows, (iii) decision-making and procedures for dispute resolutions, (iv) finance basin activities and its cooperative structures, (v) environment and sustainable management of basin resources, and (vi) engagement of stakeholder participation. The major international[6] basin organizations in Sub-Saharan Africa by region are described below, and the data used to describe the basin organizations are from Rangeley et al. (1994), ANBO, AMCOW, and GTZ (2012), and Oregon State University (2012).

B.6.1 International RBOs in the Western African Region

Gambia River Basin Development Organization (OMVG): This is an official organization and economic program that was launched in 1978 to manage Corubal, Gambia, and Geba River basins and the participant countries are Guinea and Guinea Bissau.[7] The main objective of OMVG is to promote socio-economic integration of its member's states. The specific objectives of OMVG include development of hydro-power/hydro-electricity, flood control and relief, irrigation, food security, as well as infrastructure and socio-economic development of the member states.

Mano River Union (MRU): It is an official organization established on October 3, 1973, and the participating countries are Guinea, Liberia, and Sierra Leone.

---

[6]National river basin organizations are not described here, due to limited dataset covering all Sub-Saharan countries.

[7]For the Geba River basin, OMVG includes also Senegal.

The MRU aims to manage Mana-Morro basin in order to improve living standards of participating countries.

Niger Basin Authority (*NBA*): The NBA is an official organization established in 1980 to manage Niger River basin. The NBA was born from the former Niger River Commission (RNC), established in 1964. The participating countries are Algeria, Benin, Burkina Faso, Cameroon, Chad, Guinea, Ivory Coast, Mali, Niger, Nigeria, and Sierra Leone. The aim of the Niger Basin Authority is to promote cooperation among the member countries and to ensure integrated development in all fields through development of its resources. Specifically, the NBA aims to improve water quality, hydro-power/ hydro-electricity, navigation, fishing, flood control and relief, economic development, joint management, irrigation, infrastructure development, as well as technical cooperation and assistance.

Nigeria-Niger Joint Commission for Co-operation (*NNJC*): It is an official commission established in July 18, 1990, with an objective to improve equitable sharing in the development, conservation, and use of their common water resources. Specifically, it serves as the technical body to advise the governments of the two countries on issues related to the management of Niger River basin resources.

Liptako-Gourma Integrated Authority (*Autorite de developpement integre de la region du Liptako-Gourma—ALG*): The ALG is an official organization and economic program that was established in December 3, 1970. The participating countries are Burkina Faso, Mali, and Niger. ALG's goal is to promote the integrated development of the Liptako-Gourma region in Volta River basin with a view to improving the living conditions of the population. The major management issues of ALG are to improve hydro-power/hydro-electricity, navigation, fishing, economic development, irrigation, and infrastructure development.

Organisation pour la Mise en Valeur du bassin du fleuve Senegal (*OMVS*): The OMVS is an official organization and economic program that was established in March 11, 1972. The participating countries are Mali, Mauritania, and Senegal. The OMVS was born from organization of boundary states of the Senegal River (*OERS—Organization des Etats Riverains du Sénégal*), created in 1968. The main goal of OMVS is to oversee the development of member countries through sustainable use of the Senegal River basin. The main management issues of OMVS are water quality, water quantity, hydro-power/hydro-electricity, navigation, flood control and relief, economic development, joint management, irrigation, infrastructure development, technical cooperation and assistance.[8]

B.6.2   International RBOs in the Central African Region

International Commission of Congo-Oubangui-Sangha (*CICOS*): This is an official commission composed of members from some countries that share the Congo River

---

[8]The other international river basin organizations in West Africa are Volta Basin Initiative (VBI), Volta Basin Authority (VBA), and National Agency for Niger Basin.

basin [Cameroon, Central African Republic, Republic of Congo (Brazzaville), and Democratic Republic of Congo (Kinshasa)]. This commission was launched in November 6, 1999, and it has been effectively performing its activities since November 23, 2003. The main basin issues addressed by the commission are water quality, navigation, flood control, and relief, as well as infrastructure development.

Lake Chad Basin Commission (*LCBC*): It is an official commission established in May 22, 1964. The participating countries are Cameroon, Central African Republic,[9] Chad, Niger, and Nigeria. The LCBC was established to manage the basin and to resolve disputes that might arise over the lake and its resources. The management issues of the LCB are water quality, water quantity, navigation, fishing, economic development, irrigation, infrastructure development, technical cooperation and assistance, border issues, among others.

B.6.3   International RBOs in the Eastern African Region

Nile Basin Initiative (*NBI*): The NBI is an official organization and economic program established in 1999. The participating countries are Burundi, Central African Republic, Egypt, Eritrea, Ethiopia, Democratic Republic Congo (Kinshasa), Sudan, Tanzania, Uganda, Kenya, and Rwanda. The NBI was born from the Technical Cooperation Committee for the Promotion of the Development and Environmental Protection of the Nile Basin (TECCONILE) established in 1993. The main goals of NBI are to enhance partnership, promote economic development, and fight poverty throughout the sustainable use of basin resources. Its vision is to achieve sustainable socio-economic development through the equitable utilization of Nile River basin resources.

Organization for the Management and Development of the Kagera River Basin (*Portion of Nile Basin*): It is an official organization established in February 5, 1978, and the participating countries are Burundi, Rwanda, Tanzania, and Uganda. The main issues that have been addressed under this organization is hydropower infrastructure development.

The Permanent Joint Technical Committee (*PJTC*): This is an official committee that was established in 1959. The participating countries are Egypt and Sudan. The main goals of PJTC are to implement Nile Waters Treaty Agreement of 1959, signed by the two countries to jointly manage Nile River water resources (mainly water quantity management).

Lake Victoria Fisheries Organization: It is an official organization and environmental program established in June 30, 1994. The participating countries are Kenya, Tanzania, and Uganda. The objectives of the organization are to improve cooperation among the participating countries in matters regarding Lake Victoria; harmonize national measures for the sustainable utilization of the living resources

---

[9]The Central African Republic was admitted in 1994 and at the same time Sudan was admitted as observer.

of the lake; develop and adopt conservation and management measures to assure the health of the lake's ecosystem, and the sustainability of its living resources. The main management issues are water quality, fishing, and joint management.

The Lake Victoria Basin Commission (LVBC): Similar to Lake Victoria Fishery Organization, it is an official organization that was established in June 1, 2006, and the participating countries are Kenya, Tanzania, and Uganda. The LVBC was born from the Lake Victoria Development Programme (LVDP), and its aim is to jointly manage Lake Victoria resources and mainly water quality.[10]

B.6.4   International RBOs in the Southern African Region

Tripartite Permanent Technical Commission (TPTC): It is an official commission established in February 15, 1991, to manage the Inkomati River basin. The participant countries are South Africa, Swaziland, and Mozambique. The main objectives of the TPTC are to jointly manage basin infrastructure development, as well as to perform technical cooperation and assistance among participating countries.

Joint Water Commission-Swaziland and South Africa (JWCSSA): It is also an official commission established in March 13, 1992. The JWCSSA was established as a technical advisory commission to advise the governments of Swaziland and South Africa on water resources of common interest. The JWC is actually monitoring the activities of KOBWA on behalf of the governments of Swaziland and South Africa.

Komati Basin Water Authority (KOBWA): It is an official organization and economic program established in 1993 to manage the Inkomati River basin. The participant countries are Mozambique, South Africa, and Swaziland. The purpose of KOBWA is to implement Phase 1 of the Komati River basin development project, which comprises the design, construction, operation, and maintenance of Driekoppies Dam in South Africa, and Maguga Dam in Swaziland. Mozambique is participating in KOBWA as a downstream country that can be affected by water flows from the upstream countries (South Africa and Swaziland).

Angola Namibian Joint Commission of Cooperation (ANJCC): It is an official commission established in 1996 to manage Kunene River basin. The participant countries are Angola and Namibia. The main objectives of the ANJCC are to jointly manage basin infrastructure development, as well as to perform technical cooperation and assistance among participating countries.

Limpopo Watercourse Commission (LIMCOM): It is an official commission established in November 1, 2003. The participant countries are Botswana, Mozambique, South Africa, and Zimbabwe. The main objectives of the LIMCOM

---

[10]The other Eastern Africa international basin organizations are Lake Tanganyika Authority, Awash Basin Water Resources Administration Agency, and Juba–Shabelli Basin organization.

are to manage the Limpopo River basin resources, and facilitate the building of capacity within the four countries to manage the water resource.

Limpopo River Basin Commission (LRC): Similar to LIMCOM, it is an official commission established in 1995, and the participant countries are Botswana, Mozambique, South Africa, and Zimbabwe. Different from LIMCOM, under LRC, institutional arrangement to manage water are operating on a river-catchment basis, rather than by national boundaries. The LRC provides an appropriate institutional vehicle to guide the development in the Limpopo River basin.

Limpopo Basin Permanent Technical Committee (LBPTC): It is an official committee that was established in 1986.[11] The participating countries are Botswana, Mozambique, South Africa, and Zimbabwe. The main objective of the LBPTC is to advise the parties on issues regarding the Limpopo River basin resources.

Joint Permanent Technical Committee (JPTC): This is an official organization that was established in 1983 to make recommendations on matters concerning common interest in the Limpopo River basin. Similar to other Limpopo basin organizations, the participating countries in JPTC are Botswana, Mozambique, South Africa, and Zimbabwe.

Joint Water Commission Mozambique and South Africa (JWCMSA): It is also an official commission established in 1996. The participating countries are Mozambique and South Africa. The JWCMSA is mainly playing advisory functions on technical matters to the respective governments relating Mozambique/South Africa common rivers basins, including the Limpopo basin.

The Permanent Okavango River Basin Commission (OKACOM): It is also an official commission, established in September 15, 1994. The participating countries are Angola, Botswana, and Namibia. The OKACOM is aimed to ensure that the water resources of the Okavango River system are managed in appropriate and sustainable ways, and to foster cooperation and coordination between the three basin states: Angola, Namibia, and Botswana.

Joint Permanent Water Commission (JPWC): It is an official commission established in November 13, 1990. The participating countries are Botswana and Namibia. The main goal of JPWC is to enhance bilateral management of the Okavango River and the Kwando-Chobe-Linyati basins.

Orange-Senqu River Commission (ORASECOM): It is an official commission established in November 3, 2000.[12] The participating countries are Botswana, Lesotho, Namibia, and South Africa. The ORASECOM is the first RBO to be established in terms of the SADC Protocol on Shared Watercourse Systems with the goal to manage jointly Orange-Senqu River basin.

Lesotho Highlands Development Authority (LHDA): It is an official organization and economic program established in 1930. The participating countries are Lesotho and South Africa. Initially, the LHDA was established to implement and operate the

---

[11]The LBPTC did not function during its first ten years and a second meeting aimed to revitalize it was held in South Africa in 1995.
[12]The secretariat of ORASECOM was established in 2003.

part of Lesotho Highlands Water Project (LHWP) that falls within the borders of Lesotho. Actually, LDHA has engaged on issues related to water quantity, hydro-power and hydro-electricity, economic development, joint management, and technical cooperation and assistance.

Lesotho Highlands Water Commission (*LHWC*): The LHWC was born with the signing of the Lesotho Highlands Water Project (LHWP) treaty by the government of Lesotho and of the Republic of South Africa on the October 24, 1986. In order to implement LHWP, the Joint Permanent Technical Commission (JPTC) was established to represent the two countries. The JPCT was later renamed the Lesotho Highlands Water Commission (LHWC) with the goal to oversee the LHWP treaty.

Permanent Water Commission (*PWC*): It is an official commission established in 1992. The participating countries are Namibia and South Africa. The PWC was established to act as a technical adviser to the parties on matters relating to the development and utilization of the Orange water resources.

Joint Irrigation Authority (*JIA*): It is an official organization and economic program that was established in 1992. The participating countries are Namibia and South Africa. The main goal of JIA is to administer the existing irrigation scheme along the riverbanks under the auspices of the PWC.

Zambezi Watercourse Commission (*ZAMCOM*): It is an official commission created in July 13, 2004. The participating countries are Angola, Congo, Democratic Republic of Congo (Kinshasa), Malawi, Mozambique, Tanzania, Botswana, Namibia, Zambia, and Zimbabwe. The ZAMCOM is composed of three organs: a council of ministers, a technical committee, and a secretariat, drawn from all eight countries. The secretariat advises member countries on planning, utilization, protection, and conservation issues around the Zambezi River. The major management issues are mediating disputes among participating countries.

Zambezi River Authority (*ZRA*): Like ZAMCOM, it is an official organization and economic program established in 1987. The participating countries are Zambia and Zimbabwe. The ZRA council is governed by four ministers (two from Zambia and the other two from Zimbabwe). ZRA's mission is to cooperatively manage and develop an integrated and sustainable management of the Zambezi River water resources in order to supply quality water, hydrological and environmental services for the maximum socio-economic benefits to Zambia, Zimbabwe and the other countries sharing the Zambezi River basin. ZRA's management issues are water quality, economic development, joint management, technical cooperation and assistance.[13]

---

[13]Another international basin organization in Southern Africa is the Inco-Maputo Watercourse Commission.

References

African Network of Basin Organizations (ANBO), African Ministers Council on Water (AMCOW) and GTZ. (2012). *Source book on African River Basin Organization.* Available at http://www.inbo-news.org/IMG/pdf/AWRB_Source_Book-2.pdf. Accessed in December 2012.

Oregon State University. (2012). *Program in water conflict management and transformation.* Available at http://www.transboundarywaters.orst.edu/research/RBO/RBO_Africa.html. Accessed in December 2012.

Rangeley, R., Thiam, B. M., Andersen, R. A., & Lyle, C. A. (1994). *International river basin organizations in Sub-Saharan Africa.* World Bank Technical Paper #250, Washington D.C.

United Nations Economic Commission for Africa (UNECA). (2000). Transboundary River/Lake Basin Water Development in Africa: Prospects, Problems, and Achievements. 75 p.

Annex C
Original Variables in the Dataset and Construction of Additional Variables

Name of the variable	Definition	Categories
1. barea	Area of river basin in square km	
2. ptotal	Total population in the river basin	
3. %rural	Percentage rural population in the river basin	
4. precipation	Annual precipitation/rainfall in mm	1 = 100–200 mm, 2 = 300–400 mm, 3 = 500–600 mm, 4 = 700–800 mm, 5 = 900–100, 6 = 1000–1100, 7 = 1200–1300, 8 = 1400–1500, 9 = 1600–1700, 10 = 1800–1900, 11 = 2000–2100, 12 = 2200–2300, 13 = 2400–2500, 14 = 2600–2700, 15 = 2800–2900
4. evapotransp	Annual evapotranspiration in mm	1 = 100–200 mm, 2 = 300–400 mm, 3 = 500–600 mm, 4 = 700–800 mm, 5 = 900–100, 6 = 1000–1100, 7 = 1200–1300, 8 = 1400–1500, 9 = 1600–1700, 10 = 1800–1900, 11 = 2000–2100, 12 = 2200–2300, 13 = 2400–2500, 14 = 2600–2700, 15 = 2800–2900

(continued)

© The International Bank for Reconstruction and Development/The World Bank 2016   81
J. Mutondo et al., *Water Governance Decentralization in Sub-Saharan Africa*,
SpringerBriefs in Water Science and Technology, DOI 10.1007/978-3-319-29422-3

(continued)

Name of the variable	Definition	Categories
5. wresources	River basin water resources in million cubic meters p/y	
6. countriesshare	Number of countries sharing river basin	
7. iyeadecentr	Period over which decentralization occurred in years	
8. iyearrbo	Year of creation of river basin	
9. iobjectwaterconflict	Water conflict as RBO objective	0 = No, 1 = Yes
9. iobjectflood	Flood control as RBO objective	0 = No, 1 = Yes
9. iobjectwaterscarcity	Water scarcity as RBO objective	0 = No, 1 = Yes
9. iobjectothers1,2,3	Other objective	0 = n/a, 1 = pollution, 2 = water resources management, 3 = water quality, 4 = hydropower, 5 = planning, 6 = stabilization of aquifer, 7 = conservation, 8 = water allocation/distribution, 9 = development schemes, 10 = public awareness, 11 = resource evaluation, 12 = maintenance, 13 = water management education, 14 = hydrological work, 15 = sanitation and water supply, 16 = watershed conservation, 17 = improve efficiency, 18 = navigation, 19 = flood control, 20 = water scarcity, 21 = water conflicts, 22 = water utilization, 23 = recreation, 24 = dam safety, 25 = river administration
10. ifloodscale	Measurement of success against objectives	1 = 0–19 %, 2 = 20–39 %, 3 = 40–59 %, 4 = 60–79 %, 5 = 80–100 %

(continued)

(continued)

Name of the variable	Definition	Categories
10. iwaterscarcescale	Measurement of success against objectives	1 = 0–19 %, 2 = 20–39 %, 3 = 40–59 %, 4 = 60–79 %, 5 = 80–100 %
10. iwaterconflictscale	Measurement of success against objectives	1 = 0–19 %, 2 = 20–39 %, 3 = 40–59 %, 4 = 60–79 %, 5 = 80–100 %
10. iother1scale	Measurement of success against objectives	1 = 0–19 %, 2 = 20–39 %, 3 = 40–59 %, 4 = 60–79 %, 5 = 80–100 %
10. iother2scale	Measurement of success against objectives	1 = 0–19 %, 2 = 20–39 %, 3 = 40–59 %, 4 = 60–79 %, 5 = 80–100 %
11. ibody	Governing body of river basin organization	0 = "n/a", 1 = "Federal", 2 = "State Authority", 3 = "State owned company", 4 = "Regional Authority", 5 = "Regional Board/Council/Committee", 6 = 3 and 5
12. igover-body-selct	Selection process of governing body of the river basin—nominated	1 = 'n/a', 2 = 'Federal Government', 3 = 'State', 4 = 'Local Government', 5 = 'Users'
12. igover-body-selct	Selection process of governing body of the river basin—appointed	1 = 'n/a', 2 = 'Federal Government', 3 = 'State', 4 = 'Local Government', 5 = 'Users'
12. igover-body-selct	Selection process of governing body of the river basin—designated	1 = 'n/a', 2 = 'Federal Government', 3 = 'State', 4 = 'Local Government', 5 = 'Users'
14. icreationrbo	Method of RBO creation	0 = "n/a", 1 = "Bottom-up", 2 = Top-Down
15. iinstdismantled	Institutions dismantled in decentralization process	0 = n/a, 1 = ministry/department of Water, 2 = irrigation boards, 3 = regional water authority, 4 = local authority, 5 = river boards, 6 = administration court, 7 = UDAH
16. iinewinstitution	New institutions that had to be created in decentralization process	0 = n/a, 1 = ministry/department of water, 2 = irrigation boards, 3 = regional water authority, 4 = local authority, 5 = RBO/water user

(continued)

(continued)

Name of the variable	Definition	Categories
		associations/catchment council
17. icostdecentinstitutions	Cost of the decentralization process	0 = none, 1 = low, 2 = 2, 3 = 3, 4 = 4, 5 = high
18. iforumsyesno	Do forums exist for hearing disputes	0 = no, 1 = yes
19. iissuesresolved	main types of disputes/issues that usually need resolving	0 = n/a, 1 = water quality, 2 = waste disposal, 3 = deforestation, 4 = erosion, 5 = agricultural practices, 6 = basin infrastructure, 7 = ground water pollution, 8 = floods, 9 = water allocation, 10 = siltation, 11 = water use/legal/illegal, 12 = all, 13 = 1–2–5
20. iwaterassociations	Degree of involvement of water user associations	0 = n/a, 1 = 0 %, 2 = 25 %, 3 = 50 %, 4 = 75 %, 5 = 100 %
20. iwaterassociationsyesno	Have water user associations been established	0 = no, 1 = yes
21. itypesinfrustcanal	Quantity of canals in the basin	
Before		
25. indprobbfloods	Level of flooding problems before establishment of RBO	1 = no response, 2 = no problem, 3 = some problem, 4 = severe problem
25. indprobbwaterscarcity	Level of water scarcity problems before establishment of RBO	1 = no response, 2 = no problem, 3 = some problem, 4 = severe problem
25. indprobbenvquality	Level of environmental quality problems before establishment of RBO	1 = no response, 2 = no problem, 3 = some problem, 4 = severe problem
25. indprobbwaterconflicts	Level of water conflict problems before establishment of RBO	1 = no response, 2 = no problem, 3 = some problem, 4 = severe problem
25. indprobblanddegrad	Level of land degradation problems before establishment of RBO	1 = no response, 2 = no problem, 3 = some problem, 4 = severe problem
25. indprobbdevelpissues	Level of problems with development issues before establishment of RBO	1 = no response, 2 = no problem, 3 = some problem, 4 = severe problem

(continued)

(continued)

Name of the variable	Definition	Categories
25. othername	Other problems (before and after) the establishment of RBO	0 = n/a, 1 = water mgt issues and authority crises, 2 = Env. awareness, 3 = organization, 4 = hydropower, 5 = water supply, 6 = drought
25. indprobbother	Level of other problems before establishment of RBO	1 = no response, 2 = no problem, 3 = some problem, 4 = severe problem
After		
25. indprobafloods	Level of flooding problems after establishment of RBO	−1 = situation worsened, 0 = situation the same, 1 = situation improved
25. indprobbwaterscarcity	Level of water scarcity problems after establishment of RBO	−1 = situation worsened, 0 = situation the same, 1 = situation improved
25. indprobbenvquality	Level of environmental quality problems after establishment of RBO	−1 = situation worsened, 0 = situation the same, 1 = situation improved
25. indprobbwaterconflicts	Level of water conflict problems after establishment of RBO	−1 = situation worsened, 0 = situation the same, 1 = situation improved
25. indprobblanddegrad	Level of land degradation problems after establishment of RBO	−1 = situation worsened, 0 = situation the same, 1 = situation improved
25. indprobbdevelpissues	Level of problems with development issues after establishment of RBO	−1 = situation worsened, 0 = situation the same, 1 = situation improved
25. indprobbother	Level of other problems after establishment of RBO	−1 = situation worsened, 0 = situation the same, 1 = situation improved
26. iadmblocal	Percentage of water administration decision making at local level before RBO	1 = 0–19 %, 2 = 20–39 %, 3 = 40–59 %, 4 = 60–79 %, 5 = 80–100 %
26. iadmbbasin	Percentage of water administration decision making at basin level before RBO	1 = 0–19 %, 2 = 20–39 %, 3 = 40–59 %, 4 = 60–79 %, 5 = 80–100 %
26. iadmbstate	Percentage of water administration decision making at state level before RBO	1 = 0–19 %, 2 = 20–39 %, 3 = 40–59 %, 4 = 60–79 %, 5 = 80–100 %

(continued)

(continued)

Name of the variable	Definition	Categories
26. iadmbgov	Percentage of water administration decision making at government level RBO	1 = 0–19 %, 2 = 20–39 %, 3 = 40–59 %, 4 = 60–79 %, 5 = 80–100 %
26. ifinblocal	Percentage of infrastructure financing decision making at the local level before RBO	1 = 0–19 %, 2 = 20–39 %, 3 = 40–59 %, 4 = 60–79 %, 5 = 80–100 %
26. ifinbbasin	Percentage of infrastructure financing decision making at the basin level before RBO	1 = 0–19 %, 2 = 20–39 %, 3 = 40–59 %, 4 = 60–79 %, 5 = 80–100 %
26. ifinbstate	Percentage of infrastructure financing decision making at the state level before RBO	1 = 0–19 %, 2 = 20–39 %, 3 = 40–59 %, 4 = 60–79 %, 5 = 80–100 %
26. ifinbgov	Percentage of infrastructure financing decision making at the government level before RBO	1 = 0–19 %, 2 = 20–39 %, 3 = 40–59 %, 4 = 60–79 %, 5 = 80–100 %
26. ienfblocal	Percentage of water quality enforcement decision making at the local level before RBO	1 = 0–19 %, 2 = 20–39 %, 3 = 40–59 %, 4 = 60–79 %, 5 = 80–100 %
26. ienfbbasin	Percentage of water quality enforcement decision making at the basin level before RBO	1 = 0–19 %, 2 = 20–39 %, 3 = 40–59 %, 4 = 60–79 %, 5 = 80–100 %
26. ienfbstate	Percentage of water quality enforcement decision making at the state level before RBO	1 = 0–19 %, 2 = 20–39 %, 3 = 40–59 %, 4 = 60–79 %, 5 = 80–100 %
26. ienfbgov	Percentage of water quality enforcement decision making at the government level before RBO	1 = 0–19 %, 2 = 20–39 %, 3 = 40–59 %, 4 = 60–79 %, 5 = 80–100 %
26. istdsblocal	Percentage of the setting of water quality standards decision making at the local level before RBO	1 = 0–19 %, 2 = 20–39 %, 3 = 40–59 %, 4 = 60–79 %, 5 = 80–100 %
26. istdsbbasin	Percentage of the setting of water quality standards decision making at the basin level before RBO	1 = 0–19 %, 2 = 20–39 %, 3 = 40–59 %, 4 = 60–79 %, 5 = 80–100 %

(continued)

(continued)

Name of the variable	Definition	Categories
26. istdsbstate	Percentage of the setting of water quality standards decision making at the state level before RBO	1 = 0–19 %, 2 = 20–39 %, 3 = 40–59 %, 4 = 60–79 %, 5 = 80–100 %
26. istdsbgov	Percentage of the setting of water quality standards decision making at the government level before RBO	1 = 0–19 %, 2 = 20–39 %, 3 = 40–59 %, 4 = 60–79 %, 5 = 80–100 %
26. iotherblocal26	Percentage of decision making for other responsibilities at the local level before the creation of RBO	1 = 0–19 %, 2 = 20–39 %, 3 = 40–59 %, 4 = 60–79 %, 5 = 80–100 %
26. iotherbbasin26	Percentage of decision making for other responsibilities at the basin level before the creation of RBO	1 = 0–19 %, 2 = 20–39 %, 3 = 40–59 %, 4 = 60–79 %, 5 = 80–100 %
26. iotherbstate26	Percentage of decision making for other responsibilities at the state level before the creation of RBO	1 = 0–19 %, 2 = 20–39 %, 3 = 40–59 %, 4 = 60–79 %, 5 = 80–100 %
26. iotherbgov26	Percentage of decision making for other responsibilities at the government level before the creation of RBO	1 = 0–19 %, 2 = 20–39 %, 3 = 40–59 %, 4 = 60–79 %, 5 = 80–100 %
26. iadmalocal	Percentage of water administration decision making at the local level after the creation of RBO	1 = 0–19 %, 2 = 20–39 %, 3 = 40–59 %, 4 = 60–79 %, 5 = 80–100 %
26. iadmabasin	Percentage of water administration decision making at the basin level after the creation of RBO	1 = 0–19%, 2 = 20–39 %, 3 = 40–59 %, 4 = 60–79 %, 5 = 80–100 %
26. iadmastate	Percentage of water administration decision making at the state level after the creation of RBO	1 = 0–19 %, 2 = 20–39 %, 3 = 40–59 %, 4 = 60–79 %, 5 = 80–100 %
26. iadmagov	Percentage of water administration decision making at the government level after the creation of RBO	1 = 0–19 %, 2 = 20–39 %, 3 = 40–59 %, 4 = 60–79 %, 5 = 80–100 %

(continued)

(continued)

Name of the variable	Definition	Categories
26. ifinalocal	Percentage of water administration decision making at the local level after the creation of RBO	1 = 0–19 %, 2 = 20–39 %, 3 = 40–59 %, 4 = 60–79 %, 5 = 80–100 %
26. ifinabasin	Percentage of infrastructure financing decision making at the basin level after the creation of RBO	1 = 0–19 %, 2 = 20–39 %, 3 = 40–59 %, 4 = 60–79 %, 5 = 80–100 %
26. ifinastate	Percentage of infrastructure financing decision making at the state level after the creation of RBO	1 = 0–19 %, 2 = 20–39 %, 3 = 40–59 %, 4 = 60–79 %, 5 = 80–100 %
26. ifinagov	Percentage of infrastructure financing decision making at the government level after the creation of RBO	1 = 0–19 %, 2 = 20–39 %, 3 = 40–59 %, 4 = 60–79 %, 5 = 80–100 %
26. ienfalocal	Percentage of water quality enforcement decision making at the local level after the creation of RBO	1 = 0–19 %, 2 = 20–39 %, 3 = 40–59 %, 4 = 60–79 %, 5 = 80–100 %
26. ienfabasin	Percentage of water quality enforcement decision making at the basin level after the creation of RBO	1 = 0–19 %, 2 = 20–39 %, 3 = 40–59 %, 4 = 60–79 %, 5 = 80–100 %
26. ienfastate	Percentage of water quality enforcement decision making at the state level after the creation of RBO	1 = 0–19 %, 2 = 20–39 %, 3 = 40–59 %, 4 = 60–79 %, 5 = 80–100 %
26. ienfagov	Percentage of water quality enforcement decision making at the government level after the creation of RBO	1 = 0–19 %, 2 = 20–39 %, 3 = 40–59 %, 4 = 60–79 %, 5 = 80–100 %
26. istdsalocal	Percentage of the setting of water quality standards decision making at the local level after the creation of RBO	1 = 0–19 %, 2 = 20–39 %, 3 = 40–59 %, 4 = 60–79 %, 5 = 80–100 %

(continued)

(continued)

Name of the variable	Definition	Categories
26. istdsabasin	Percentage of the setting of water quality standards decision making at the basin level after the creation of RBO	1 = 0–19 %, 2 = 20–39 %, 3 = 40–59 %, 4 = 60–79 %, 5 = 80–100 %
26. istdsastate	Percentage of the setting of water quality standards decision making at the state level after the creation of RBO	1 = 0–19 %, 2 = 20–39 %, 3 = 40–59 %, 4 = 60–79 %, 5 = 80–100 %
26. istdsagov	Percentage of decision making on setting of water quality standards at the government level after creation of RBO	1 = 0–19 %, 2 = 20–39 %, 3 = 40–59 %, 4 = 60–79 %, 5 = 80–100 %
26. iothername	Other responsibilities	1 = quality objectives, 2 = O and M, 3 = management, 4 = planning, 5 = water supply
26. iotheralocal	Percentage of the decision making for other responsibilities at the local level after the creation of RBO	1 = 0–19 %, 2 = 20–39 %, 3 = 40–59 %, 4 = 60–79 %, 5 = 80–100 %
26. iotherabasin	Percentage of the decision making for other responsibilities at the basin level after the creation of RBO	1 = 0–19 %, 2 = 20–39 %, 3 = 40–59 %, 4 = 60–79 %, 5 = 80–100 %
26. iotherastate	Percentage of the decision making for other responsibilities at the state level after the creation of RBO	1 = 0–19 %, 2 = 20–39 %, 3 = 40–59 %, 4 = 60–79 %, 5 = 80–100 %
26. iotheragov	Percentage of the decision making for other responsibilities at the government level after the creation of RBO	1 = 0–19 %, 2 = 20–39 %, 3 = 40–59 %, 4 = 60–79 %, 5 = 80–100 %
28. wrmibresponsiblerigths	Responsibility for awarding water rights before RBO existence	0 = n/a, 1 = Federal, 2 = National Agency, 3 = State/Provincial, 4 = Regional Organization, 5 = National Agency, 6 = River Basin Organization

(continued)

(continued)

Name of the variable	Definition	Categories
29. wrmibresponsibleallocation	Responsibility for water allocation before RBO existence	0 = n/a, 1 = Federal, 2 = National Agency, 3 = State/Provincial, 4 = Regional Organization, 5 = Local Government, 6 = River Basin Organization
30. wrmibresponsiblemodfore	Responsibility for modeling and forecasting water availability before RBO existence	0 = n/a, 1 = Federal, 2 = National Agency, 3 = State/Provincial, 4 = Regional Organization, 5 = Local Government, 6 = River Basin Organization
31. wrmibresponsiblemonit	Responsibility for monitoring and enforcement of water quality before RBO existence	0 = n/a, 1 − Federal, 2 = National Agency, 3 = State/Provincial, 4 = Regional Organization, 5 = Local Government, 6 = River Basin Organization
32. wrmiaresponsibletariff	Responsibility for collecting tariffs after RBO existence	0 = n/a, 1 = Federal, 2 = National Agency, 3 = State/Provincial, 4 = Regional Organization, 5 = Local Government, 6 = River Basin Organization
27. wrmibwatertypes	Water rights after RBO existence	0 = none, 1 = permanent rights, 2 = long-term use concession (>10 years), 3 = short-term use concession (<10 years), 4 = permanent transferable, 5 = permanent non-transferable
28. wrmibresponsiblerigths	Responsibility for awarding water rights after RBO existence	0 = n/a, 1 = Federal, 2 = National Agency, 3 = State/Provincial, 4 = Regional Organization, 5 = Local Government, 6 = River Basin Org
29. wrmibresponsibleallocation	Responsibility for water allocation after RBO existence	0 = n/a, 1 = Federal, 2 = National Agency, 3 = State/Provincial, 4 = Regional Organization, 5 = Local Government, 6 = River Basin Org

(continued)

(continued)

Name of the variable	Definition	Categories
30. wrmibresponsiblemodfore	Responsibility for modeling and forecasting water availability after RBO existence	0 = n/a, 1 = Federal, 2 = National Agency, 3 = State/Provincial, 4 = Regional Organization, 5 = Local Government, 6 = River Basin Org
31. wrmibresponsiblemonit	Responsibility for monitoring and enforcement of water quality after RBO existence	0 = n/a, 1 = Federal, 2 = National Agency, 3 = State/Provincial, 4 = Regional Organization, 5 = Local Government, 6 = River Basin Org
32. wrmiaresponsibletariff	Responsibility for collecting tariffs after RBO existence	0 = n/a, 1 = Federal, 2 = National Agency, 3 = State/Provincial, 4 = Regional Organization, 5 = Local Government, 6 = River Basin Org
53. part-intl-bsn-treaty	River basin part of an international basin	0 = no, 1 = yes
54. flow-var-flact-overtime	Does water flow in basin fluctuate across the year	0 = no, 1 = yes
55. res-dist-equal-bfor-decentr	River resources equitably distributed	0 = no, 1 = yes
56. bfor-ben-2-gov	Who benefited most before rbo	1 = federal government, 2 = local leaders, 3 = commercial farmers, 4 = small farmers
57. res-dist-equal-aftr-decentr	Basin resources equitably distributed after RBO	0 = no, 1 = yes
58. ftr-ben-2-gov	Who benefited most after rbo	1 = federal government, 2 = local leaders, 3 = commercial farmers, 4 = small farmers

Annex D
Revised River Basin Organization (RBO) Questionnaire

Dear Survey Respondent[14]:

This survey is part of a research project that tries to assess in which way the creation of River Basin Organizations (RBO) leads to the decentralization of water resources management to other (lower levels) of decision-making. The research project also tries to assess in which way the creation of RBOs leads to improved water resources management results.

The specific information (in the box below) regarding each individual basin will be kept in confidentiality not to allow identification of the River Basin Organization.

The results of the research effort will be made publicly available and, hopefully, help in the continent-wide effort to bring about sustainable integrated water resources management.

If you find you do not have enough space to fill out the questionnaire, you can expand the sections in this Word document or provide annexing sheets.

Your collaboration in this effort is highly appreciated.

---

[14]This questionnaire is the result of adaptation made in the questionnaire developed by Dinar et al. (2005)

© The International Bank for Reconstruction and Development/The World Bank 2016
J. Mutondo et al., *Water Governance Decentralization in Sub-Saharan Africa*,
SpringerBriefs in Water Science and Technology, DOI 10.1007/978-3-319-29422-3

1. RIVER BASIN IDENTIFICATION

Basin:_____Country:_____

RBO Name:_____

RBO Address:_____

_____

_____

Contact Person: _____Telephone:_____Fax::_____

2. DECENTRALIZATION PROCESS

Part A: Laws, Acts and Decrees

2.1. Has the country developed and enacted water related laws, decrees, acts, etc. that have influenced the management of water resources in the country? *1. Yes; 2. No*

2.2. If yes in question 2.1., have the local people contributed to the development of water related issues (laws, decrees, acts, etc.): *1. Yes; 2. No*

2.3. If yes to question 2.2., who was more active in crafting the rules?

1. Politicians; 2. Government officials; 3. Traditional structure and local people

4. Other_____; 5. Other_____

2.4. How often these rules are broken by the local people?

1. Never broken; 2. Seldom broken; 3.Regularly broken; 4. Not followed at all.

2.5. In your opinion, did the present water laws contribute to decentralization of water resource management? *1. Yes; 2. No.* Why? _____

_____
_____
_____

2.6. What are the main objectives of the water law in the country? _____

_____
_____

2.7. To date, are those objectives attained?

1. Not at all; 2. 25% attained; 3. 50% attained; 4. 75% attained; 5. 100% attained

2.8. Period (years) that the decentralization took place in the country_____

Part B: Institutions

3.1. What was the Year that the River Basin Organization was created_____

3.2. What was the type of devolution of the River Basin Organization Creation?

1. Top-down; 2. Bottom-up; 3. Both

3.3. Who came up with the first idea of forming the River Basin Organization?

_____

_____

_____

_____

3.4. Who created the River Basin Organization?*1. Government; 2. Private sector; 3.Civil society; 4.Local community; 5.NGOs6. Other_____*

3.5. Have the local people contributed to the development of the River Basin Organization? *1. Yes; 2. No*

3.6. If yes to question 3.15, who was more active in creating the River Basin Organization?
1. Politicians; 2. Government officials; 3. Traditional structure and local people
4. Other_____; 5. Other_____

3.7. Can you explain in detail the River Basin Organization creation process?

_____

_____

_____

_____

3.8. Describe the existing organizations that had to be dismantled in the decentralization process at national
level_____

_____

_____

3.9. Describe the new organizations that had to be created in the decentralization process including their role and administrative power in the country_____

_____

_____

_____

_____

_____

3.10. What are the existing organizations at river basin level that had to be dismantled in the decentralization process?_____

_____

_____

3.11. What are the new organizations at river basin level that had to be created in the decentralization process? _____

_____

_____

3.12. What were the costs of creating organizations due to decentralization process?
a. None b. Low cost c. Medium cost d. High cost

3.13. In developing the river basin organization, what are the difficulties that have been encountered in the process if any?_____

_____

_____

_____

_____

3.14. What are the main objectives of the River Basin Organization?
*1. Flood control; 2. Water scarcity; 3.Water conflicts; 4.Assuring water quality; 5. Other*_____

3.15. To date are those objectives attained?

Flood Control	Water Scarcity	Water Conflicts	Assuring Water Quality	Other
O N/A	O N/A	O N/A	O N/A	O N/A
O 1 (0% success)	O 1 (0% success)	O 1 (0% success)	O 1 (0% success)	O 1 (0% success)
O 2 (25% success)	O 2 (25% success)	O 2 (25% success)	O 2 (25% success)	O 2 (25% success)
O 3 (50% success)	O 3 (50% success)	O 3 (50% success)	O 3 (50% success)	O 3 (50% success)
O 4 (75% success)	O 4 (75% success)	O 4 (75% success)	O 4 (75% success)	O 4 (75% success)
O 5 (100% success)	O 5 (100% success)	O 5 (100% success)	O 5 (100% success)	O 5 (100% success)

3.16. Can you please provide the River Basin Organization organigram?

_____

_____

_____

3.17. Explain the roles of each element of the organigram

_____

_____

_____

3.18. Can you please provide the composition of governing body of the river basin organization including the type of stakeholders (water users) that they represent as well as the level of education?

3.19. Explain the process by which the Governing Body of the River Basin Organization was selected

_____

_____

_____

_____

Name	Type of water user	Education

3.20. Does the River Basin Organization have human capacity to manage water resource at basin level? *1. Yes; 2. No.*

3.21. Are there capacity building programs for the River Basin Organization's stakeholders? *1. Yes; 2. No.* If yes, explain the types of capacity building (training courses, seminars, study tours, etc.)_____

_____

_____

_____

3.22. Explain the laws of the land and decrees that govern the River Basin Organization. Please provide your answer using chronological order._____

_____

_____

_____

Part C: Finance

3.23. Do you measure your basin's revenues? *1. Yes; 2. No If no, please go to question 3.26.*

3.24. If yes in question 3.23, please indicate the basin's yearly revenues and the basin population in the past five years.

Year	Revenues	River Basin Population
2010		
2009		
2008		
2007		
2006		

3.25. What is the value of the river basin's revenues by sector?

Sectors	Revenues
Agriculture	
Forestry	
Industry	
Other (name_____)	
Other (name_____)	

3.26. What is the value of water Tariffs for different water users (if possible provide rates for various major users):

Water Users	Water tariffs
Irrigation	
Industry	
Domestic	
Other_____	
Other_____	
Other_____	
Other_____	

3.27. Can you indicate the percentage of users paying tariffs for the different water users? Indicate in table below using the following choices of percentage of water users paying tariffs: *1. Not applicable; 2. 0%; 3.25%; 4.50%; 5. 75%; 6. 100%.*

User group	Percentage who pay
Irrigation	
Industry	
Domestic	
Other_____	
Other_____	
Other_____	
Other_____	

3.28. Which percentage of the tariff payments stays in the basin and which percentage goes to other destinations? Which destinations?

3.28a. Percentage of tariffs staying in the Basin: *1. Not applicable; 2. 0%; 3.25%; 4.50%; 5.75%; 6. 100%.*

3.28b.Percentage of tariffs going to other Destinations: *1. Not applicable; 2. 0%;*

3. 25%; 4. 50%; 5. 75%; 6. 100%.

3.28c.What are the destinations of water tariff _____

3.29. Extent/activities of private sector involvement in basin investments (e.g. water supply, water treatment, reservoir construction, basin infrastructure maintenance): Percent Private Involvement: _____ *(1. Not applicable 2. 0%   3. 25% 4. 50% 5. 75% 6. 100%)*

3.30. What is the annual budget of the river basin organization? _____

3.31. What are the major sources and their contribution for the annual budget?

Sources	Percentage (0-100%)
Government	
Privatesector (name_____)	
NGOs (name_____)	
Stakeholders at River Basin	
Other (name_____)	
Other (name_____)	

3.32. What is the distribution of the annual budget in percentage among different activities at River Basin?

Activities	Percentage (0-100%)
Investment	
Development	
Water quality	
Capacity building and meetings	
Other (name_____)	
Other (name_____)	

3.33. Does the River Basin Organization have the necessary authority/independence in managing water resources? *1. Yes; 2. No.*
Why_____
_____
_____
_____
____

3.34. Are some of the decisions made by the River Basin Organization delayed by the government? *1. Yes; 2. No.*

3.35. If yes to question 3.34, how do you rate the impact of these delays on service delivery? *1. None; 2. Moderate; 3. Severe*

Part D: Information sharing

3.36. How often the River Basin Organization call for a meeting?*1. Never; 2. When need rise; 3. Twice a year; 4.Quarterly; 5.Monthly6. Other_____*

3.37. Can you rate the participation of stakeholders at the meeting? Percentage of members attending the meeting (0-100%)_____

3.38. What types of issues are frequently discussed on these meetings?*1. Politics and non water issues; 2. Some water issues; 3. Purely important water issues4. Other_____; 5. Other_____*

3.39. What is the percentage of time allocated to each of the following issues at these meetings?

Meeting issue	Percentage(%)
1. Politics and non water issues_____	
2. Some water issues	_____
3. Purely important water issues	_____

3.40. What are the other forms of information sharing among stakeholders (annual reports, websites, radio, etc.) and explain their effectiveness in communicating to all stakeholders

_____
_____
_____
_____

Part E: Disputes and their Resolution

3.41. Are there forums to hear disputes, how many and which ones?_____
_____
_____

3.42. What are the main types of disputes/issues that usually need to be resolved?_____
_____
_____
_____

3.43. How often these conflicts rise? *1. Never; 2. Rarely; 3.Often; 4.Very often.*

3.44. What are the challenges faced by the River Basin Organization in resolving the conflicts?_____

_____

_____

_____

4. DECENTRALIZATION PERFORMANCE

4.1. Indicators of problems before and after establishment of the RBO. Please check all that apply in the table bellow for each water resource problem at river basin before and after de establishment of RBO using the following choices: 1. *No response; 2. No problem; 3. Some problem; 4. Severe problem.*

Water resource problem at the River basin	Before	After
Water scarcity		
Floods		
Environmental quality		
Land degradation (erosion, salinity, etc.)		
Water conflicts (water allocation, etc.)		
Water storage		
River ecology		
Other (specify)		
Other (specify)		

4.2. Describe the major water resource problems at the river basin before and after the decentralization process in terms of occurrence and consequences._____

_____

_____

_____

4.3. Responsibilities for decision making **before** and **after** the creation of the RBO. Please indicate the share of decision making of different levels of governance (municipal, basin, provincial and national) for the areas (water administration, etc.) indicated in table below **before** and **after** the establishment of RBO using the following choices of share (in %) in decision making: *1. Not applicable; 2. 0%; 3.25%; 4. 50%; 5. 75%; 6. 100%*

Responsibility for:	Before the creation of the RBO				After the creation of the RBO			
	% at local level (e.g municipality)	% at Basin level	% at state/ provin-cial gov. level	% at national gov. level	% at local level (e.g municipality)	% at Basin level	% at state/provincial gov. level	% at national gov. level
Water Administration								
Infrastructure Financing								
Water quality enforcement								

Setting water quality standards									
Other (please explain)									

4.4. Water Resource Management Instruments: Compare the situation **before** and **after** the existence of the RBO:

	Before RBO	After RBO
Existence of water right types (e.g. concessions, permanent rights, short-term rights qualitative or quantitative):	O None O Permanent Rights O Long-Term Use Concession (more than 10 years) O Short-Term Use Concession (less than 10 years) O Permanent Transferable O Permanent Non-Transferable Other:	O None O Permanent Rights O Long-Term Use Concession (more than 10 years) O Short-Term Use Concession (less than 10 years) O Permanent Transferable O Permanent Non-Transferable Other:
Who is responsible for awarding water rights:	O N/A O Federal O State/Provincial O Local Government O Regional Organization O National Agency O River Basin Organization	O N/A O Federal O State/Provincial O Local Government O Regional Organization O National Agency O River Basin Organization

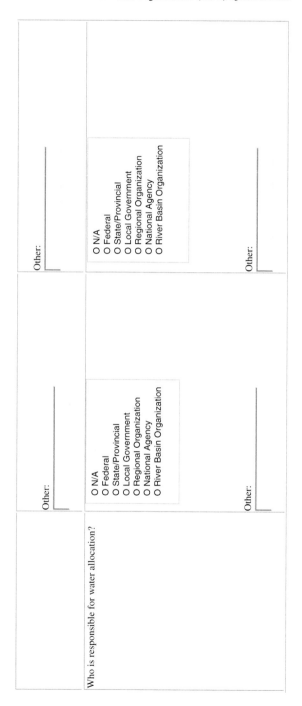

Who is responsible for water allocation?

Other:

O N/A
O Federal
O State/Provincial
O Local Government
O Regional Organization
O National Agency
O River Basin Organization

Other:

O N/A
O Federal
O State/Provincial
O Local Government
O Regional Organization
O National Agency
O River Basin Organization

Other:

Other:

Who is responsible for modeling and forecasting water availability?

O N/A
O Federal
O State/Provincial
O Local Government
O Regional Organization
O National Agency
O River Basin Organization

Other: |___

Who is responsible for monitoring and enforcement of water quality?

O N/A
O Federal
O State/Provincial
O Local Government
O Regional Organization
O National Agency
O River Basin Organization

O N/A
O Federal
O State/Provincial
O Local Government
O Regional Organization
O National Agency
O River Basin Organization

Other: |___

O N/A
O Federal
O State/Provincial
O Local Government
O Regional Organization
O National Agency
O River Basin Organization

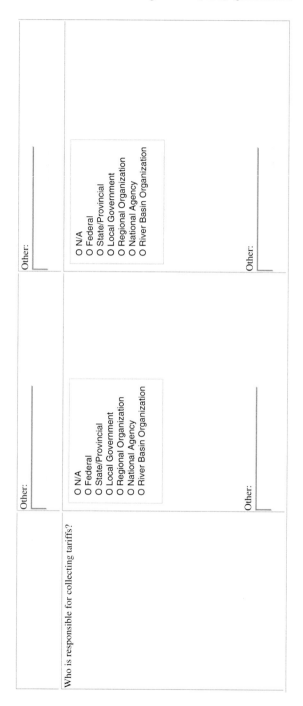

Who is responsible for collecting tariffs?

Other:

Other:

O N/A
O Federal
O State/Provincial
O Local Government
O Regional Organization
O National Agency
O River Basin Organization

O N/A
O Federal
O State/Provincial
O Local Government
O Regional Organization
O National Agency
O River Basin Organization

Other:

Other:

4.5. Describe the reduction in loss of production and productivity due to water scarcity or flooding **before** and **after** the decentralization process?_____

_____
_____
_____
_____

_____

4.6. Quantify and describe disputes regarding water allocation or water quality **before** and **after** the creation of the River Basin Organization

_____
_____
_____
_____

5. BASINS COMPARISONS

5.1. In your opinion, are there some characteristics about this river basin that make it different from other basins in the country? *1. Yes; 2. No*

5.2. If yes in question 5.1, what are these characteristics and can you please mention the strengths and weaknesses of these characteristics?

Strengths:

Weaknesses:

5.3. Is the river basin in question part of an international basin that is subject to an existing treaty or an international RBO? 1. Yes; 2. No

5.4. In your opinion does the water flow in the basin highly fluctuate across years? 1. Yes; 2. No

5.5. In your opinion does the river basin resources uniformly distributed before the decentralization process? 1. Yes; 2. No

5.6. If no in question 5.5, who benefited more from river basin resources before the decentralization process? 1. Government, 2. Local leaders, 3. Commercial Farmers; 4. Smallholder farmers, 5. Other_____

5.7. In your opinion does the river basin resources uniformly distributed after the decentralization process? 1. Yes; 2. No

5.8. If no in question 5.7, who benefited more from river basin resources after the decentralization process? 1. Government, 2. Local leaders, 3. Commercial Farmers; 4. Smallholder farmers, 5. Other_____

5.9. Any comments clarifications including annexed material you think may be of value?

Annex E
Result Tables

See Tables E.1, E.2, E.3, E.4, E.5, E.6, E.7, E.8, E.9, E.10, E.11, E.12, E.13 and E.14.

Table E.1 Studied variables and their derived impact on decentralization process and performance in the three studied catchments

Variables	Possible impact on decentralization process and performance		
	Mzingwane (Zimbabwe)	Inkomati (South Africa)	Limpopo (Mozambique)
Contextual factors and initial conditions			
Level of economic development	▼▼[a]	▲	▼▼
Distribution of resources among basin stakeholders	▼	▼	▼
Stakeholders managerial skills	▲	▲▼	▼
Characteristics of decentralization process			
Composition of catchment boards and degree of stakeholders participation	▼▼	▲	▲▼
Stakeholders involvement in decentralization process	▼	▲	▼
Type of devolution of the decentralization process	▼	▲	▼
Central government/basin-level relationships and capacities			
Source of river basin budget	▲▼	▼	NA
Percentage of water tariffs remaining at the basin	▲▼	▼▼	▼▼
Level of management authority given to basin stakeholders	▲▼	NA	▼
Configuration of basin-level institutional arrangements			
Presence of basin-level governance institutions	▲	▲	▲

(continued)

© The International Bank for Reconstruction and Development/The World Bank 2016   111
J. Mutondo et al., *Water Governance Decentralization in Sub-Saharan Africa*,
SpringerBriefs in Water Science and Technology, DOI 10.1007/978-3-319-29422-3

Table E.1 (continued)

Variables	Possible impact on decentralization process and performance		
	Mzingwane (Zimbabwe)	Inkomati (South Africa)	Limpopo (Mozambique)
Information sharing	▲▼	▲▼	▲
Mechanism for conflict resolution	▲▼	▲	▼

[a]▲▲ = highly positive impact; ▲ = positive impact; ▼▼ = highly negative impact; ▼ = negative impact; ▲▼ = contrasted impact

Table E.2 Decentralization performance according to respondents in the three studied catchments

Decentralization performance	Mzingwane (Zimbabwe)	Inkomati (South Africa)	Limpopo (Mozambique)
Level of accomplishment of river basin objectives	▲▼[a]	▲▼	▼
Improvement of problems related to river basin stressed resources after decentralization	▲▼	▲▼	▼
Introduction of water permits	▲	▲	▲

[a]▲▲ = very good performance; ▲ = good performance; ▼▼ = very bad performance; ▼ = bad performance; ▲▼ = contrasted performance

Table E.3 Decentralization process

Independent var.	Dependent var.		
	WUAs involvement	RBO created	Institutions dismantled
Budget per capita	NI	NI	NI
Creation bottom-up	+	+	+
Disputes over allocation	−	+	NI
Governing body	NI	NI	NI
International treaty	+	+	+
Political cost	+	+	+
Relative water scarcity	NI	+	+
Share of surface water	NI	NI	+
Water flow fluctuates	NI	NI	+
WUA involvement	NI	NI	NI
Years decentralization	−	NI	NI

NI Not included

Table E.4 Decentralization performance

Independent var.	Dependent var.	
	Success over objectives	Problems after decentralization
Budget per capita	NI	+
Creation bottom-up		+
Disputes over allocation	NI	NI
Governing body	+	NI
Institutions dismantled	NI	NI
International treaty	+	NI
Political cost	−	−
RBO created	NI	NI
Relative water scarcity	NI	NI
Share of SW	+/−	NI
Water flow fluctuates	−	NI
WUA involvement	NI	NI
Years decentralization	+	NI

NI Not included

Table E.5 Initial set of identified river basins in SSA by region

Region	Number of reported river basins
Southern Africa	34
West Africa	30
Central Africa	14
East Africa	21
Total	99

Source ANBOAMCOW and GTZ, 2012

Table E.6 List of identified river basins in Sub-Saharan Africa

Number	River basin	Country
Southern Africa		
1	Berg	South Africa
2	Cuanza	Angola
3	Cuvelai/Etosha	Namibia, Angola
4	Fish	Namibia
5	Groot	South Africa
6	Ihosy	Madagascar
7	Kafue	Zambia
8	Inkomati	South Africa, Swaziland, Mozambique
9	Kuiseb	Namibia
10	Kunene	Angola (as Cunene), Namibia, Botswana

(continued)

Table E.6 (continued)

Number	River basin	Country
11	Kwando	Namibia
12	Limpopo	Mozambique, South Africa, Zimbabwe, Botswana
13	Buzi	Mozambique, Zimbabwe
14	Luangwa	Zambia
15	Licungo	Mozambique
16	Ligonha	Mozambique
17	Lurio	Mozambique
18	Messalo	Mozambique
19	Mangoky	Madagascar
20	Mania	Madagascar
21	Maputo/Usutu/Pangola	South Africa, Swaziland, Mozambique
22	Molopo	Botswana, South Africa
23	Okavango	Botswana, Angola, Namibia
24	Onilahy	Madagascar
25	Orange	South Africa, Namibia, Lesotho
26	Pungwe	Mozambique, Zimbabwe
27	Shangani	Zimbabwe
28	Tugela	South Africa
29	Vaal	South Africa
30	Zambezi	Angola, Zambia, Namibia, Zimbabwe, Mozambique
31	Savi/Sabi	Mozambique, Zimbabwe
32	Rovuma	Mozambique, Tanzania, Malawi
33	Umbeluzi	Mozambique, Swaziland
34	ReVive	Zambia, Namibia, Zimbabwe
Central West Africa		
1	Bandama	Côte d'Ivoire
2	Cavally	Liberia, Côte d'Ivoire
3	Cestos	Liberia, Côte d'Ivoire
4	Komoe	Côte d'Ivoire, Burkina Faso, Ghana, Mali
5	Gambia	Gambia, Senegal, Guinea
6	Niger	Nigeria, Benin, Niger, Mali, Guinea
7	Oueme	Benin
8	Saint Paul	Liberia
9	Sanaga	Cameroon
10	Akpa	Cameroon
11	Atui	Mauritania, Western Sahara
12	Sankarani	Mali
13	Sassandra	Côte d'Ivoire, Guinea
14	Tano	Ghana, Côte d'Ivoire
15	Corubal	Guinea, Guinea Bissau

(continued)

Table E.6 (continued)

Number	River basin	Country
16	Senegal	Mauritania, Mali, Senegal
17	St. Jone (Africa)	Liberia, Guinea
18	Geba	Guinea-Bissau, Senegal, Guinea
19	Great Scarcies	Guinea, Sierra Leone
20	Little Scarcies	Sierra Leone, Guinea
21	Loffa	Liberia, Guinea
22	Mana-Morro	Liberia, Siera Leone
23	Mbe	Gabone, Equatoria Guinea
24	Moa	Sierra Leone, Guinea
25	Mono	Togo, Benin
26	Volta	Ghana, Burkina Faso
27	Bia	Côte d'Ivoire, Ghana
28	Cross	Nigeria, Cameroon
29	Utamboni	Gabon, Equatorial Guinea
30	Benue	Nigeria
Central África		
1	Logone–Chari	(Central African Republic)
2	Kwango	Congo
3	Kasai	Congo
4	Lualaba	Congo
5	Lomami	Congo
6	Chiloango	Democratic Republic of the Congo
7	Uele–Ubangi	Democratic Republic of the Congo
8	Mbomou–Ubangi	Democratic Republic of the Congo
9	Gabon	
10	Kouilou-Niari	Congo
11	Mbini/Benito	Equatorial Guinea
12	Ntem	Equatorial Guinea, Cameroon, Gabon
13	Nyanga	Gabon
14	Ogooué	Gabon
East África		
1	Awash	Ethiopia, Djibouti, Somalia
2	Jubba	Somalia
3	Dawa	Ethiopia
4	Gebele	Ethiopia
5	Kerio	Kenya
6	Lotagipi Swamp	Kenya, Sudan
7	Baraka	Eritrea, Sudan
8	Gash	Eritrea, Sudan, Etiopia
9	Lake Natron	Tanzania, Republic of Kenya

(continued)

Table E.6 (continued)

Number	River basin	Country
10	Lake Turkana	Ethiopia, Kenya
11	Umba	Tanzania, Republic of Kenya
12	Mara	Kenya, Tanzania
13	Omo	Ethiopia
14	Nile	Sudan, Ethiopia
15	Lake Chad	Chad, Niger
16	Atbarah	Sudan, Ethiopia
17	Blue Nile	Sudan, Ethiopia
18	Didessa R	Ethiopia
19	Mountain Nile	Sudan
20	Bahr el Zeraf	Sudan
21	White Nile	Sudan

Table E.7 Distribution of decentralization efforts in various regions of SSA

Country	Basins with decentralization undertaken	Basins with decentralization in progress	Basins with no decentralization	Basin with no information about decentralization
Southern Africa Region				
Angola			7	
Botswana			4	
Lesotho			1	
Madagascar			4	
Mozambique[a]	13			
Namibia		10		
South Africa	2	17		
Swaziland	1	2		
Zambia			3	
Zimbabwe	7			
Subtotal	23	29	19	0
West Africa Region				
Ivory Coast				1
Benin				1
Liberia				1
Cameroon				2
Ghana			4	
Guinée				1
Mali				1
Mauritania				1
Nigeria				1
Senegal				1

(continued)

Table E.7 (continued)

Country	Basins with decentralization undertaken	Basins with decentralization in progress	Basins with no decentralization	Basin with no information about decentralization
Subtotal	0	0	4	10
Central African Republic				1
DR Congo			4	4
Equatorial Guinea				1
Gabon				2
Subtotal	0	0	4	8
East Africa Region				
Ethiopia				4
Kenya		5		
Malawi			1	
Sudan				4
Tanzania	9			
Uganda			1	
Subtotal	9	5	2	8
Central Africa Region				
Central African Republic				1
Democratic Republic Congo			4	4
Equatorial Guinea				1
Gabon			1	1
Subtotal	0	0	6	8
Total	32	34	29	26

[a]Mozambican respondents to our survey indicated that RBOs in that country are established. Compared to the level of development of the RBOs of other African countries, it would probably be more correct to put Mozambican RBOs in the second column, where water decentralization process is "in progress". However, to reflect precisely the survey results, we decided to leave the Mozambican RBOs in the first column

Source Modified from PEGASYS (2013)

Table E.8 Details about the basins included in our analysis

	Basins with decentralization undertaken	Basins with decentralization in progress	Basins in sample	Names of basins included
Mozambique	13		5	Limpopo, Inkomati, Buzi, Save, Pungwe
Kenya		5	1	Lake Victoria
South Africa	2	17	10	Breede-Overberg, Incomati, Olifants/Letaba, Middle Vaal, Upper Orange, Crocodile, Usuthu, Thukela, Mvoti, Limpopo
Swaziland	1	2	2	Komati, Usuthu
Zimbabwe	7		6	Gwayi, Limpopo, Save, Sanyati, Manyame, Mazowe
Tanzania	9		3	Rufuji, Wami/Ruvu, Internal Drainage
Total in sample	30	26	27	
Total in region (Table E.2)	30	36	N/A	N/A

Note While some similar basin names can be found in different countries, each represent a different RBO, with no physical or institutional interaction between these RBOs

Table E.9 The final RBOs included in the analysis

River basin organization	Country
Lake Victoria	Kenya
AraSul Limpopo	Mozambique
Ara Centro Buzi	Mozambique
AraCentorPungue	Mozambique
Ara Centro Save	Mozambique
AraSulInkomati	Mozambique
Komati River Basin Authority	Swaziland
Usuthu River Basin Authority	Swaziland
Breede Overberg Catchment Management Agency	South Africa
Inkomati Usuthu Catchment Management Agency	South Africa
Crocodile West Marico Proto Catchment Management Agency	South Africa
Upper Orange Proto Catchment Management Agency	South Africa
Mvoti to Umzimkulu Proto Catchment Management Agency	South Africa
Middle Vaal Proto Catchment Management Agency	South Africa
Tukela Proto Catchment Management Agency	South Africa
Usutu to Mhaltuze Proto Catchment Management Agency	South Africa

(continued)

Table E.9 (continued)

River basin organization	Country
Olifants Proto Catchment Management Agency	South Africa
Limpopo Proto Catchment Management Agency	South Africa
Rufiji Basin Water Board	Tanzania
WamiRuvu Basin Water Board	Tanzania
Internal Drainage Basin Water Board	Tanzania
Gwayi Catchment Council	Zimbabwe
Manyame Catchment Council	Zimbabwe
Mazowe Catchment Council	Zimbabwe
Mzingwana Catchment Council	Zimbabwe
Sanyati Catchment Council	Zimbabwe
Save Catchment Council	Zimbabwe

Source PEGASYS (2013: 33)

Table E.10 Descriptive statistics of variables included in the analysis

Variable	Obs	Mean	Std. Dev.	Min	Max
River basin part of an international basin	25	0.68	0.4760	0	1
Does water flow in basin fluctuates across the year	25	0.76	0.4358	0	1
River basin resources equitably distributed	25	0.16	0.3741	0	1
Budget percapita	17	6.6131	15.7686	0.1785	66.4250
Forum to solve dispute	23	1.0869	0.4170	0	2
Governing body	22	4	1.661	1	6
Method of creation	27	1.5925	0.5007	1	2
Creation bottom-up	27	0.4074	0.5007	0	1
Creation top-down	27	0.5925	0.5007	0	1
Existence of political cost	25	3.56	1.3868	0	5
Relative water scarcity	17	0.5230	0.3308	0.0864	1.5
Share surface water	23	4.4781	0.9472	1	5
Water Users Association involvement	24	1.6666	1.007	1	5
Year of creation	18	1999	7.3163	1979	2009
Years of decentralization	23	9.4782	6.4938	2	30
RBO created	25	0.800	0.4082	0	1
Institutions dismantled	17	0.5882	0.5072	0	1
Disputes over quality	23	0.5217	0.5107	0	1
Disputes over allocation	23	0.3478	0.4869	0	1
Problems before decentralization (PC variable)	15	2.41e−09	0.9482	−2.3690	2.4236
Problems after the decentralization (PC variable)	10	−1.34e−08	0.9765	−1.1872	1.3384
Success over objectives (redefined)	16	5.4375	1.6720	3	9

Table E.11 Decision making in water management at various levels before and after decentralization

Activity	Before	After	t-Statistic
Water administration			
Local	2.235	2.692	0.8785
Basin	1.611	3.733	6.0498***
State	2.875	3.125	0.3369
Central government	3.950	2.533	−2.7947***
Infrastructure financing			
Local	1.917	2.400	0.9659
Basin	1.286	2.714	2.4019**
State	3.222	3.125	−0.1453
Central government	4.714	4.667	−0.1166
Water quality enforcement			
Local	1.500	1.800	0.7069
Basin	1.529	3.273	3.7063***
State	2.750	2.500	−0.4229
Central government	4.000	3.286	−1.8609*
Setting water quality standards			
Local	1.200	1.000	−0.5311
Basin	1.333	2.333	2.3094**
State	2.083	2.714	0.9073
Central government	4.600	4.571	−0.1031

Note ***$p < 0.01$; **$p < 0.05$; *$p < 0.10$

Table E.12 Changes in severity of various water management issue between before and after decentralization

Problem item	Before	After	t-Statistic
Floods	0.9545	0.7222	1.5396[+]
Water scarcity	1.0952	0.4705	3.6246***
Environmental quality	1.1052	0.2666	3.5794***
Water conflicts	1.3888	0.2666	4.5825***
Land degradation	1.0500	0.7500	1.6771*
Development issues	1.3333	0.6153	3.5257**

Note ***$p < 0.01$; **$p < 0.05$; *$p < 0.10$; [+]$p < 0.15$. We included also coefficients with level of significance of 15 % to accommodate results that are influenced by the small number of observations

Table E.13 Estimated features of the decentralization process

Estimation procedure	OLS	OLS	LPM	LPM	LPM
Explanatory variable	WUAs involvement	WUAs involvement	RBO created	RBO created	Institutions dismantled
Political cost	1.10711 (4.41)***	1.10686 (5.00)***	0.4717735 (3.32)**	0.5731967 (4.79)***	0.2062154 (4.04)**
Creation bottom-up	−1.033681 (2.19)*	−1.108916 (2.61)**	−0.249556 (3.36)**	−0.307502 (4.90)***	−0.085916 (7.99)**
Years decentralization	−0.367126 (5.11)***	−0.363617 (5.73)***			
Disputes over allocation	−1.030815 (2.23)**	−0.846964 (1.98)*	0.4499993 (3.22)**	0.7309282 (4.67)***	
Relative water scarcity			0.901773 (3.16)**	1.160028 (4.84)***	0.9306318 (14.08)***
Share of surface water					0.1589505 (13.30)***
International treaty		0.7457297 (1.78)+		0.2751419 (1.99)+	0.1759502 (5.20)**
Water flow fluctuates					0.7785227 (11.71)***
Constant	1.67017 3.03	1.063595 (1.75)+	0.8078305 (2.97)**	0.5119992 (2.15)*	−0.789900 (9.10)**
Number of obs	16	14	11	10	9
F-test	7.42	6.83	5.18	8.4	285.08
Prob > F	0.0038	0.0091	0.0377	0.0302	0.0035
R-squared	0.7295	0.8103	0.7754	0.9131	0.9988
Adj R-squared	0.6312	0.6918	0.6257	0.8045	0.9953

Note Absolute value of t-statistics in parenthesis. +significant at 15 %, *significant at 10 %, **significant at 5 %, ***significant at 1 %

Table E.14 Estimated decentralization performance equations

Estimation procedure	OLS	OLS	OLS	OLS
Dependent variable	Success over objectives	Success over objectives	Success over objectives	Problems after decentralization
Share of surface water	0.5967261 (3.39)**	0.5868282 (10.37)***	0.5931021 (9.74)***	
Years decentralization	0.1928462 (3.18)**	0.1395445 (6.31)***	0.1450607 (6.21)***	
Political cost	−1.104221 (7.38)***	−1.019237 (20.25)***	−1.009395 (16.80)***	−1.071558 (8.50)***
Governing body	0.9838797 (6.18)***	0.954158 (18.72)***	0.9483496 (15.83)***	

(continued)

Table E.14 (continued)

Estimation procedure	OLS	OLS	OLS	OLS
Dependent variable	Success over objectives	Success over objectives	Success over objectives	Problems after decentralization
Creation bottom-up				7.296772 (8.04)***
Budget per capita				0.9797866 (7.79)***
Water flow fluctuates		−0.108023 (0.75)		
International treaty			−0.012094 (0.10)	
Constant	1.608739 (1.2)	2.123604 (4.37)**	1.96945 (4.02)**	−3.63149 (5.31)***
Number of obs	10	9	9	7
F-test	33.71	276.39	233.62	26.84
Prob > F	0.0008	0.0003	0.0004	0.0114
R-squared	0.9642	0.9978	0.9974	0.9641
Adj R-squared	0.9356	0.9942	0.9932	0.9282

Note Absolute value of t-statistics in parenthesis. +significant at 15 %, *significant at 10 %, **significant at 5 %, ***significant at 1 %

Annex F
Figures

See Figs. F.1 and F.2.

© The International Bank for Reconstruction and Development/The World Bank 2016   123
J. Mutondo et al., *Water Governance Decentralization in Sub-Saharan Africa*,
SpringerBriefs in Water Science and Technology, DOI 10.1007/978-3-319-29422-3

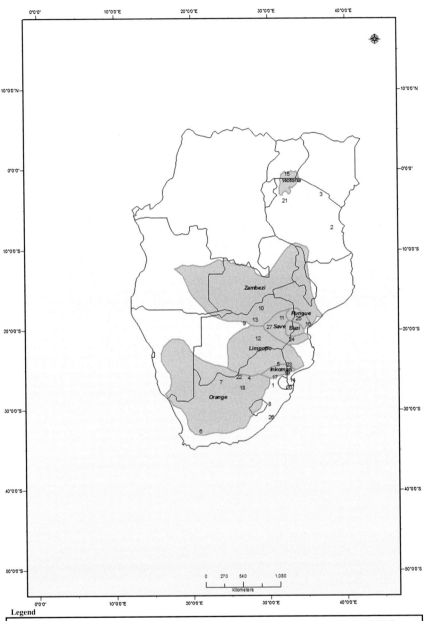

Legend

1.Usuthu River Authority; 2.Rufiji Basin Water Board; 3.Wami Ruvu Water Board; 4.Crocodiles West M.P.C. Agency; 5.Oliffants Proto C.M. Agency; 6. Breed Overberg C.M. Agency; 7.Middle Vaal P.C.M. Agency; 8.Tugela P.C.M. Agency; 9.Gwayi Catchment Council; 10. Manyame Catchment Council; 11.Mazowe Catchment Council; 12.Mzingwana Catchment Council; 13.Sanyati Catchment Council; 14.ARA Sul – Limpopo; 15.Lake Victoria; 16.ARA Centro – Save; 17.Inkomati Suthu C.M. Agency; 18.Upper Orange P.C.M. Agency; 19.Komati River Basin Authority; 20.Usuthu River Basin Authority; 21.Internal Drainage B. Water Board; 22.Limpopo P.C.C.M. Agency; 23.ARA Sul – Inkomati; 24.ARA Centro – Buzi; 25. ARA Centro – Pungue; 26.Mvoti to Umzikulu P.C. Agency; and 27.Save C. Council

Fig. F.1 Geographical location of the interviewed RBOs. *Source* Adapted from DNTF data, 2011

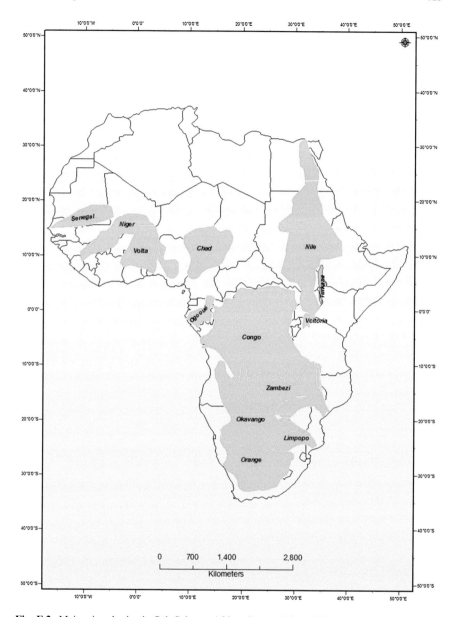

Fig. F.2 Major river basins in Sub-Saharan Africa. *Source* Adapted from DNTF data, 2011

Printed in the United States
By Bookmasters